Silent Strategists

Harding, Denby, and the U.S. Navy's Trans-Pacific Offensive, World War II

Revised Edition

Manley R. Irwin

University Press of America,® Inc.
Lanham • Boulder • New York • Toronto • Plymouth, UK

Copyright © 2013 by University Press of America,® Inc.
4501 Forbes Boulevard, Suite 200, Lanham, Maryland 20706
UPA Aquisitions Department (301) 459-3366

10 Thornbury Road, Plymouth PL6 7PP, United Kingdom

All rights reserved
Printed in the United States of America
British Library Cataloguing in Publication Information Available

Library of Congress Control Number: 2013932908
ISBN: 978-0-7618-6101-0 (paperbound : alk. paper)—ISBN: 978-0-7618-6102-7 (electronic)

♾™ The paper used in this publication meets the minimum requirements of American
National Standard for Information Sciences Permanence of Paper for Printed Library
Materials, ANSI/NISO Z39.48-1992.

For Ellen, Katie, Patricia, Mary, Marjorie, Douglas

Contents

Preface

Any study is a work in progress and this edition is no exception. The present update is the result of reexamining several dockets in U.S. naval archives. They include the Division of Fleet Training, the Bureau of Construction and Repair, the U.S. Base Force, the Secretaries Council minutes, the Annual Reports, Commander in Chief of the U.S. Fleet and the History of the U.S. Naval Research Laboratory.

I owe a debt of gratitude to the following individuals, Fred Hall, George Findell, Professors Louis Knight, Steve Hardy, Douglas Wheeler, Pete Chinburg, Homer Bechtell, Ernie Nichols. Robert Raymond kept the flame alive. Dr. Douglas Irwin and Dr. Marjorie Rose provided critical and insightful feedback. Dr. John Kraus has remained ever supportive, patient and understanding. Patricia Irwin and Mary Taylor listened to naval history with remarkable forbearance. Professor John and Sally Freear endured discussions of naval history that bordered on unusual punishment.

The UNH library has been most supportive. I owe a debt of gratitude to Debbie Watson and her colleagues at the library's reference desk; to Jan Salas, interlibrary loan services; to Linda Johnson, and Catherine Horrigan, government document librarians; to Jean Putnam and Sara Stinson, the library's loan desk.

Mark Mollan, U.S. Naval and Maritime Archiver, was invaluable in making documents available for my research. I thank Jeff Bridges, Library of Congress, for providing unpublished photos from the Harding era. Special thanks are due to Alice Juda for navigating me through the Naval War College's library, Newport, Rhode Island.

Sinthy Kounlasa was indispensible in converting illegible drafts into a readable print out. My wife, Doris, endured endless recitations on naval history. She could tell the story more concisely than its author. Her support and counsel made *Silent Strategists II* possible.

I bear responsibility for all mistakes, large and small.

Introduction

The bill of particulars against the Harding administration's naval policies is familiar and long standing. The administration succumbed to a naval agreement that scrapped some thirty capital ships. The administration accepted a ten year holiday on battleship construction. The administration confined its Pacific base to Pearl Harbor, Hawaii, ceding control of the western Pacific to the Imperial Japanese Navy. The administration imposed draconian budget cuts while encouraging corporate access to U.S. naval oil reserves – dissipating a scarce, national resource. The fact that Harding's Secretary of the Interior, Albert Fall, served a one year prison term for fraud, together with a Senate vote against the stewardship of Edwin Denby, Secretary of the Navy, captures the essence of an administration not merely anti-navy but anti-public as well. Warren Harding and Edwin Denby have yet to recover from the scandal of Teapot Dome.

This study revisits the naval policies of this controversial administration. We do so for the following reasons. First, conventional studies have tended to concentrate on Harding's domestic and foreign agenda—taxes, tariffs, oil reserves, reparations, the World Court. Few have addressed the administration's naval policy as a study unto itself.

Second, historians have tended to overlook the Presidential papers of Warren Harding, an oversight that is quite understandable. Following the President's death in August 1923 Harding's personal papers were presumed lost or otherwise destroyed. Six years later, however, the Hoover administration discovered boxes in the basement of White House. The White House sent Harding's presidential papers to the Harding family in Ohio. Their contents remained unavailable until the early 1960's when the Ohio Historical Foundation duplicated Harding's papers on microfilm. Today they are available at over 100 university libraries.

Third, many of the naval documents, letters, reports, and correspondence remained classified until the 1960's and 1970's and thus beyond the reach of historians. Some documents, in fact, were not released until the 1980's. Now available, this trove of information provides additional insight that into the policies of the President and his navy secretary.

Fourth, the papers of Edwin Denby, located at the Burton Historical Collection at the Detroit Public Library and the Bentley Collection at the University of Michigan, have generally been overlooked by students of naval history. Combined with U.S. Marine Corps archives of the early 1920's, these records detail the evolution of amphibious operations in the period between the two world wars.

The following chapters follow an organizational matrix. One side of the matrix includes administrative oversight, a balanced fleet, amphibious warfare, logistics ashore and afloat, material sourcing, industrial mobilization, and inter-agency coordination. A word about each.

On an administrative level the President stood as the navy's commander in chief. Subject to Senate approval, a navy secretary carried out the President's naval policies. Congress, in 1915, created the Office of the Chief of Naval Operations led by the navy's ranking admiral (CNO). Over time Congress had organized the navy department into a series of functional bureaus, engineering, ordnance, construction and repair, personnel (navigation) yards and docks, et al. Each bureau was administered by a chief holding the rank of Rear Admiral, subject to Senate confirmation. Individual bureaus received their funding from congress. Once appropriated the President, the Navy Secretary, the Chief of Naval Operations were banned from transferring monies from one bureau to another.

The U.S. Fleet included an array of combatant ships, carriers, cruisers, battleships, destroyers, submarines. A balanced fleet presumed a tonnage distribution of warship classes. In operation the fleet adhered to a set of tactical guidelines known as fleet doctrine.

The Marine Corps, an arm of the navy, became intrigued with amphibious warfare in the 1920's. Given the vast expanse of the Pacific, the Marines proposed to seize an enemy-held island, converting that island into a launching pad for future base acquisition. The success of any joint operation depended on the coordination of the fleet, the Marines, and the Army units.

Logistic supply touched on two broad categories, ashore and afloat. The former included navy yards, stations, factories, arsenals, and oil reserves, an assembly termed the shore establishment. Logistics afloat included a collection of cargo, repair ships, distilling ships, oilers, tugs, barges, and tenders, sometimes called the train.

The matter of navy material acquisition reduced itself to a make-or-buy decision. In the case of a make option, the shore establishment constructed and fabricated vessels, guns, ship hulls, propulsion machinery and associated components via its in-house factories and arsenals. In the case of a buy option, the navy turned to commercial shipping yards residing in the nation's industrial sector. The latter choice, in today's parlance, is known as outsourcing.

In the event of a national emergency the navy supplemented its navy yard complex by turning to industrial mobilized firms in the nation's private sector. Under those conditions the navy submitted a list of material preferences that translated into production schedules.

Intergovernmental operations addressed the matter of agency coordination, an attempt to reach some element of cooperation among and between the armed services. Although the locus of coordinator rested with the president congress often determined whether the army and navy adopt a formal or informal unified command system.

We next examine the other matrix side, the policies of the executive and legislative branch of government; namely, Wilson, chapter one; Harding, chapter two; Coolidge-Hoover, chapter three; the first two terms of Franklin Roosevelt; chapter four. Chapter five outlines the U.S. Fleet's offensive action in the central Pacific, commencing in the fall of 1943. Chapter six highlights the application of modern management principles instituted by Warren Harding and Edwin Denby. Chapter seven details the destruction of Edwin Denby, specifically the controversy over naval oil reserve policy.

Chapter eight, the Denby Interval, places the secretary's management and administrative reforms between the two world wars. Here we suggest that Denby's policies anticipated the navy's administrative imperatives of the second World War in the central Pacific. Put differently, Edwin Denby's naval policies embodied a remarkable blend of vision, foresight and decisiveness. He also appreciated the latent potential of scientific management. Neither comprehending nor understanding the reach of those policies the U.S Senate rendered its ultimate judgment. They voted to remove the secretary from office. The Senate was successful. On March10, 1924, Denby stepped down as navy secretary.

Chapter One

The Wilson Administration, 1913-1921

ADMINISTRATIVE OVERSIGHT

Woodrow Wilson offered the American electorate a compelling resume. A political scientist by training, Wilson had been a distinguished scholar, lawyer and professor at Princeton University. He rose through the ranks of academia and became president of the university. He entered state politics, was elected governor of New Jersey, and later nominated as candidate for the U.S. Presidency at the 1912 Democratic Convention. That year, the Republican Party split between supporters of former President Theodore Roosevelt and those who favored President William Howard Taft. Although a combined Republican vote exceeded the Democrats, the Republican fracture gave Wilson a plurality. He assumed the office of the presidency in March of 1913.

Wilson would serve two terms as president, and though he suffered a massive stroke in 1919, he contemplated a third run in 1920. The Democrats selected James M. Cox, Governor of Ohio, and Franklin D. Roosevelt of New York to run against Warren Harding and Calvin Coolidge. Wilson retired from public life and died in early 1924.

Wilson's Cabinet, William Jennings Bryan, Secretary of State; Albert S. Burleson, Postmaster General; Josephus Daniels, Secretary of the Navy, reflected Democratic Party's the southern populist tradition. Wilson's 1912 election platform, a "New Freedom", rested on his conviction that U.S. corporations dominated the economy, restricted consumer choice, misapplied U.S. antitrust laws, and sought tariff protection to insulate domestic earnings from overseas competition.

Wilson, confining his advisors to a select group both in and outside the government, practiced a somewhat eclectic management persona. On one side he sent General John J. Pershing to France with instructions to aid the Allied cause, essentially leaving Pershing alone to conduct the war as he saw fit. Unlike Lincoln, Wilson rarely second-guessed or badgered his generals. On the other hand, the President regarded his cabinet as subordinates assigned to carry out the Wilson's domestic agenda. Reluctant to delegate, Wilson often chose to perform that function on his own, especially in matters of foreign affairs.[1]

The President was especially wary of his military advisors. Wilson regarded military meetings, absent his approval, as an infringement upon executive prerogatives. Josephus Daniels, Secretary of the Navy, was even more suspicious of the Navy's officer corps. With respect to the armed forces, Wilson and Daniels shared a common belief.[2]

Secretary of the Navy

A former newspaper publisher from North Carolina, Josephus Daniels, was critical of the nation's industrial sector, and especially antagonistic of any corporation capitalized over $100 million dollars. In his mind, corporate size personified corporate greed.

Although the public perceived U.S. Naval officers as serving the Nation's first line of defense, Daniels took exception to that view. In his mind, the Navy's officer corps constituted privileged elite. Daniels was convinced that the Navy should emulate an environment of learning and education. Only then would noncommissioned personnel be able to achieve their latent talent and individual aspiration. A battle cruiser was thus more than an artifact of war. It should be a classroom for instruction and education.[3]

Daniels's philosophy may have invited ridicule from some officers, a view reinforced by Daniels' flat hat and ribbon tie attire. The smirks disappeared when Daniels, a teetotaler, ordered that the Navy would be dry from now on. Admiral Bradley Fiske, Operations head, reminded Daniels that if naval officers were denied their table wine they might be tempted to indulge in cocaine.[4]

Over the years, Congress kept the Navy Secretary on a short administrative leash. Money proved the ultimate control mechanism. Although the bulk of the department's budget went to the Navy's material bureaus, Congress denied the Navy Secretary sufficient resources to manage the department's multimillion dollar operation. Daniels did, of course, turn to the Department's General Board for advice and counsel. The Board's agenda was all encompassing, ranging from battleship design to ship propulsion units, from fleet structure to war planning. Most Board members were scheduled for retirement. For many officers, the General Board would be their last official assignment.

Although the Board possessed an institutional memory, whether its officers were up to date on maritime matters was an open question. In any case, Congress banned the Board from issuing direct orders to the Department's bureau chief. The Board's standing remained advisory only.

Prior to the Civil War, Congress divided the Navy Department along functional bureaus or units– steam engineering, yards and docks, construction and repair, ordnance, navigation, etc. Once Congress set a bureau's budget, any Navy Secretary was prohibited from transferring funds between bureaus. In so doing, Congress denied a Secretary the very essentials of management authority. During testimony before a Congressional committee, the Secretary was reduced to the role of an interested onlooker. And Secretary's were expendable. President Theodore Roosevelt, for example, ran through five Secretaries in seven years. [5]

As a civilian, any Secretary faced impediments in running his office. Not unlike other specialties the Navy had its own peculiar jargon. Most Secretaries had to acquire knowledge of the Navy's nomenclature or at least have it translated into English.

Given its bases, districts, stations, factories, garment shops, and fleet afloat, the Navy took on the attributes of a government holding company. Grounded by tradition, bound by seniority, commanded by rank, the Navy became bottom heavy. The real authority resided, not in the Secretary's office, but rather with the U.S. House and Senate Naval Affairs Committees. It was Congress that opened yards, built cruisers, demobilized ships, promoted officers.

To that extent, the navy's table of organization was somewhat misleading. If the Department performed its mission well, Congress willingly accepted the Nation's gratitude. If, on the other hand, Navy personnel committed an embarrassing misstep Congressional accountability quickly evaporated. Instead, House and Senate Naval committees rounded up Naval officers to mete out guilt, punishment, and in some cases dismissal. Management by Congressional committee was not without its advantages. No single lawmaker was saddled with personal accountability.

If the Navy's structure appeared unwieldy, one might be tempted to argue the department's organization was simply a mirror image of congress's committee system - duplicative, uncoordinated, and prone to quarrel over jurisdictional turf. Rarely did Congress examine the operation of its own hodgepodge of overlapping committees. It was enough to know that the committee system rested on seniority and seniority defined political power.

Given the locus of Congressional authority, it would be misleading to suggest that a Navy Secretary had nothing to do. On the contrary, the Secretary's workload was overwhelmed by a blizzard of paper, personnel transfers, disciplinary cases, 10,000 civilian yard employees, officer promotions, decommissioned ships, Navy Day preparations. One Secretary captured the essence of his daily office routine when he recalled that an officer would approach the Secretary's desk, lay down a paper, point to a dotted line and say, "sign your name here."[6]

The Secretary's paper flow was of such magnitude that there were not enough hours in the day to reduce one's daily in-basket. A Secretary's workload also imposed an opportunity cost. Few Secretaries had time to view the organization in a broad context of technological change, dual use products, management effectiveness or the underlying forces altering on the nation's industrial sector. As a consequence, a civilian Secretary depended upon information garnered from the very organization he was told to manage. Absent a staff worthy of its name a Navy Secretary could easily be reduced to a cipher.

George von Meyer, President William Howard Taft's Navy Secretary, did attempt to impose a semblance of coordination over the navy's bureau system. He added an administrative tier of operations, materials, personnel and inspection, over the half dozen bureau chiefs. The chiefs thus not only ran their own bureaus, they also served as an advisory group to the Secretary. Meyer attempted to strike a balance between central oversight and decentralized autarchy.[7]

Josephus Daniels regarded Meyer's aid system as an anathema. In Daniels' view, Meyer's organizational fix had "Prussianized" the Navy, hardly a term of endearment at the time when German Troops torched Belgium's priceless Louvaan Library in their 1914 march toward Paris. Not surprisingly, Daniels began to unravel Meyer's aid appointments. In one instance, the Secretary assigned a captain to sea duty, leaving the aid slot unfilled. Nor was Daniels alone. The U.S. House and Senate refused to provide funds for Meyer's coordinating effort.

In addressing the navy's Bureau Chiefs, Josephus Daniels resorted to a divide and conquer strategy. He preferred to deal with each Bureau Chief individually. It was Daniels' way to prevent bureau chiefs from end-running the Secretary and taking any grievance to Capitol Hill.[8]

After the Civil War, Congress created the Office of Assistant Secretary to ease the Secretary's administrative burden. Although the Assistant Secretary's duties were somewhat amorphous, the Bureau Chiefs stood in opposition to the new position. The Chiefs need not have been worried. It was Congress that exercised cognizance over warship construction, shore funding monies, and bureau expenditures. And Congress stood as the financial court of last resort.

In his first term, Josephus Daniels appointed a Naval Consulting Board to advise the Secretary. Chairing the board, Thomas A. Edison, suggested that the Department formed a research unit comparable to Edison's laboratory in New Jersey. In 1916, Congress allocated $1.5 million for the construction of a Naval Research Laboratory.

Edison held a deep conviction as to the laboratory's administration make-up facility. If Naval officers were put in charge of civilian's R&D effort, Edison feared a military mindset would smother the laboratory's scientific endeavor. Apparently, Daniels was unable to decide on the laboratory's administration structure. By the end of his eight years in office, the staffing and structure of the NRL remained in limbo.

Daniels welcomed a young member of the New York legislator, Franklin D. Roosevelt, to serve as Assistant Navy Secretary. Daniels took an immediate liking to Roosevelt and would later describe their relationship as "love at first sight." They would serve together nearly eight years in the Wilson Administration.[9]

Though his office duties were ill defined, Franklin Roosevelt thrived as Assistant Secretary. He ranged over so many naval matters that at one time he confessed that he really ran the Department. Roosevelt's management style stood in contrast to Daniels, who stewed over problems and was prone to procrastination. Roosevelt, by contrast, proved to be a quick study and was action oriented. When serving as Acting Secretary in Daniels absence, Roosevelt "moved the paper."[10]

The Office of the Chief of Naval Operations

In his first year in office Woodrow Wilson faced the usual mix of domestic and foreign issues. California had banned Japanese residents from the right to own property in the state in 1913. Hearing of the state's move, Japan fired off a letter to the U.S. demanding appropriate action to be taken against California's Japanese bias.

Japan's response prompted a meeting of the Joint Army-Navy Board. Established after the Spanish-American War (1903), the Board was given authority to coordinate the war planning activities of the armed services. President Wilson learned about the board's meeting from a newspaper article. The President assembled the Cabinet and insisted that the board had met without Wilson's permission. If the Joint Board had the temerity to meet again, Wilson threatened to abolish the board forthwith. An Army officer discretely reminded the President that the Joint Board was a creature of Congress. Daniels, nevertheless, recorded in the diary that "it was thrilling to see the president excoriate against the military." Daniels celebrated his Commander-in-Chiefs' put down of a military elite.[11]

A year later Admiral Bradley Fiske, Operations aid, viewed tensions between Germany, France and Britain as ominous. Fiske questioned how an impending European war might impact the U.S. in general and the U.S. Navy in particular. As a line officer, Fiske had long concluded that the navy's bureaus were overly powerful. He was equally skeptical over the responsiveness and performance of the navy's shore establishment.

Admiral Fiske was convinced that the U.S. industrial sector could out produce any federal government agency. Fiske's conviction contradicted Daniels' belief and commitment to the navy's shore establishment. The Secretary insisted that Navy yards were efficient, productive, and responsive to the needs of the fleet. More important, the Navy yards could not be accused of generating obscene profits.

Fiske remained concerned by what he viewed as a bottom-heavy bureau structure. Taking a leaf from Secretary von Meyers, Fiske insisted that the Navy needed a coordinating body comparable to the Army's general staff system. Daniels, by contrast, regarded a staff system as a threat to civilian control of the Navy. Fiske proceeded to send Daniels' memos and reports regarding the department's overall administrative structure. Daniels listened politely but said little.

Rebuffed and frustrated, Fiske took matters into his own hands. He drafted legislation that created a naval Operations office and handed the proposal to Richmond Hobson, a member of the U.S. House of Representatives. An Annapolis graduate and a hero of the Spanish-American War, Hobson filed legislation that contemplated an Office of Naval Operations headed by a ranking admiral. Operations would be responsibility for war planning authority, the department's bureau system, and operating fleet. [12]

Secretary Daniels raced to Capitol Hill to head off Hobson's bill. Indeed, Daniels threatened to resign if Congress approved Hobson's legislation. In 1915, Congress passed a law according the CNO (Chief of Naval Operations) responsible for war planning, subject to approval by the Navy Secretary. The legislation, however, stripped the CNO of any direct authority over both the navy's bureau system and the fleet afloat. From the bureau's chief's perspective, the integrity of a decentralized organization had been preserved. From the line officer's perspective, Congress had compromised the CNO authority to carry out its overriding mission, fleet readiness.

It was not long before Admiral Fiske found himself in serious trouble. In going to Congress, Fiske had by-passed the Secretary of the Navy. Thereafter, Daniels told the admiral that "you cannot write or talk any more; you can't even say that two and two make four." Fiske was sent to the Naval War College, Newport, Rhode Island and later took retirement. Fiske's legislative end run had effectively ended his naval career. The alliance between Daniels and Capitol Hill remained undisturbed. [13]

Daniels now sought a naval officer qualified to carry on as the Navy's first Chief of Naval Operations. Several candidates turned down the post. When Daniels lamented his frustration to Woodrow Wilson, the president responded, "Get a captain." Daniels offered the CNO post to the Commandant of the Philadelphia Navy Yard, Captain William Benson. Benson accepted the position and would serve as CNO throughout the First World War.

In 1919, Admiral Benson retired from the Navy and accepted an appointment to the U.S. Shipping Board.[14]

Daniels' originally tried to ban Operations from engaging in any war planning activity at all. Congress did fund twelve staff positions for war planning activities. After Congress declared war in 1917, the CNO's staff had all it could do to meet the daily military requirements of the Atlantic fleet. Operation's staff, in fact, suffered such burn out that some officers had to be placed on sick leave.[15]

FLEET

When war erupted in August 1914, Germany faced Russia on the east and France on the west. Confronting a two front war, the German army planned a quick envelopment of the French army to effect its collapse. Once France capitulated, the German army could take its time dealing with lethargic Russian divisions in the East. German troops did get off to a fast pace in Belgium. But the Paris envelopment degenerated into a slow grind, assisted by a lethal combination of machine guns, trenches, barbed wire and howitzers. Siege warfare displaced a war of movement, and progress at the front, originally calculated in miles, was now measured in yards. Placing the fleet adjacent to the Baltic in the North Sea, Britain's Royal Navy imposed a blockade fleet to deny German export earnings, supplies, food and fuel. The British fleet had long excelled in a classic economic squeeze.

By late spring 1916, the Germany High Seas Fleet ventured from port to engage Britain's Grand Fleet off the coast of Jutland. Lasting for several days, the German fleet lost 11 ships and had 3,000 casualties. The Royal Fleet suffered a loss of 14 ships and incurred 6,000 casualties. Still, the Royal Navy claimed a victory of sorts. The naval embargo held and the German High Seas Fleet never again ventured from port, until the Armistice of November 1918.[16]

After the guns fell silent in November, the allies quarreled over the spoils of war. A key issue was how to dispose of the German Fleet. The matter was conveniently settled when German officers released their vessel's water cocks and the interned war ships sank to the bottom of Scotland's Scapa Flow.

The battle of Jutland lived on, resonating in the halls of naval colleges both in the U.S. and abroad. Jutland ascended to the rank of a classic engagement, an encounter to be studied, analyzed, and memorized by students preparing for fleet command. U.S. naval midshipmen poured over every move and countermove. If nothing else, Jutland confirmed the admonition of Captain Alfred Mahan, the U.S naval historian, that force concentration remained an essential to naval success.[17]

As Germany, France, Britain and Russia spun out of control in what some historians would call Europe's civil war, the Wilson administration adopted a posture of neutrality, abetted by two protective moats, the Pacific and the Atlantic. Britain's naval embargo, however, affronted the American belief of freedom of the sea, much to the distress of U.S. naval officers who harbored memories of the War of 1912. The U.S. maintained a neutral stance in 1914 and much of 1915. During that period, the President cautioned Secretary Josephus Daniels to hold the line on naval spending. Daniels did so.

Germany responded to Britain's blockade by imposing an embargo of its own; the submarine. German U-boats plied the Atlantic and dispatched torpedoes to merchant ships bearing supplies to the British Isles. William Jennings Bryans, Wilson's Secretary of State, had warned the President that a British Liner, the Lusitania, was thought to be carrying contraband cargo. In 1915, a Germany U-Boart sank the Lusitania with a loss of lives including American passengers. Wilson sent a stiff protest note to the German government, so strongly worded that William Jennings Bryan tendered his resignation on grounds the President had violated his own neutrality pledge.[18]

In the latter months of 1915, President Woodrow Wilson began to distance himself from a policy of non-rearmament. The President subsequently sent Congress a program calling for the construction of 156 vessels including ten battleships and six battle cruisers - a three year program estimated to cost a half billion dollars. The President wanted the U.S. fleet to exceed the tonnage of Britain's Royal Navy. Given the prickly relations between the two nations, some U.S. naval officers supported that challenge. Admiral William Benson, Chief of Naval Operations, was reputed to have told naval officers assigned to London that he was just as willing to fight Britain as Germany. Twisting the Lion's tail was always great fun.[19]

In deference to U.S. public opinion, Germany scaled back its U-boat offensive. By January 1917, however, Germany reactivated its submarine assault, recognizing that such action might trigger a war with the U.S. German military leaders gambled that their ground offensive would drive a wedge between British and French troops. France would drop out of the war forcing Britain to sue for peace. All this would take place before U.S. troops could reinforce the allies. Once again, the German army placed a premium on movement and speed.

President Wilson, in January 1917, severed relations with Germany and asked Congress for a declaration of war. On April 6th, Congress so voted. Britain rushed a delegation to Washington, D.C. The war had exhausted England's financial resources and Britain wanted to borrow one billion dollars from the U.S. New York now eclipsed London as the world's preeminent capital market.[20]

British officers briefed Admiral Sims the U.S. Navy commander assigned to London. German U-Boats, warned Britain, had reduced England's food supply to an estimated six weeks. Having dismissed the feasibility of a convoy, the Admiralty had no effective response to Germany's submarine offensive.

In London, Admiral William Sims called upon Operations to embark on a program of antisubmarine vessels. He also requested destroyers and merchant ships to convoy food stuff and war materials across the Atlantic. To supplement Britain's continental embargo, the U.S. Navy proposed a mine barrage barrier across the North Sea. British officers expressed doubt that the scheme would work. Assistant Secretary Franklin Roosevelt supported the plan and played a decisive role in pushing the barrier to completion. [21]

Germany's U-Boat offensive was so effective that the Wilson administration had to defer its 1916 battleship program and embark on a crash construction of destroyers, destroyer escorts, and submarine chasers (Eagle Boats). All had a cumulative effect. By 1918, the German submarine crisis began to ease.

In the spring of 1918, General Eric Ludendorff launched a western front offensive. The attack decimated the British Sixth Army as German troops penetrated Allied lines by miles. German troops soon outran their supplies and the offensive lost its punch. The allies then struck back. Under the direction of General Ferdinand Foch, a series of counter attacks set in motion a steady German withdrawal. On November 11, 1918, Germany agreed to a ceasefire. The Armistice had caught the U.S. by surprise. The U.S. Army had been preparing for a 1919 offensive.

Three weeks after the Armistice, President Wilson announced that the U.S. would not only resume its 1916 construction program, it would double that warship tonnage. By this time, the estimated construction cost had escalated to some three billion dollars. Wilson now prepared for peace talks in Paris. [22]

The battleship stood as the program's centerpiece; cruisers, destroyers, submarines defined by their supporting dreadnaught role. The Wilson administration was prepared to build more warships than all other nations combined. It was also the year that Britain introduced the world's first aircraft carrier. [23]

Naval Aviation

Aviation experienced giant strides during the First World War, most notably the Royal Navy. The Royal Naval Air Service (RNAS), the fleet's aviation arm, was the first to launch a plane from a moving ship, the first to launch the torpedo by plane, the first to engage in a strategic bombing raid (bombing strikes on German Zeppelin sheds), and the first to introduce the aircraft carrier. In naval aviation, the RNAS stood in a class by itself.

Assigned to the British Expeditionary Force, the Royal Flying Corps (Army) conducted reconnaissance missions in support of army ground action. General Hugh Trenchard, Douglas Haig's aviation assistant, worked hand in glove with Britain's five armies on the western front. In November 1917, the Royal Flying Corps buzzed German lines as a noise camouflage that enabled the British third Army to hide its tanks along the front. Absent pre-artillery rounds, the British attack caught German troops by surprise and the battle of Cambrai attained the distinction as the first massed tank attack in the annals of military history. The penetration by the British Third Army was of such depth that London church bells rang in celebration. The tank offensive, however, lost its staying power and in a furious counter-attack the Germans pushed the British Forces beyond their original staging area.

The Allied offensive of August 4, 1918 proved a different enterprise. The Royal Flying Corps, again in conjunction with infantry troops, participated in an operation that combined tanks, artillery, and infantry into a set battle piece. The Allies continued their ground offensive until November 1918.

Of necessity, British Army and Navy units had divided their respective aviation units during the First World War. The Royal Flying Corps was assigned to assist ground troops on the western front; the Royal Naval Air Service positioned to defend the British Isles. In the summer of 1917, German bombers evaded both air units and struck London's Charing Cross railroad station, inflicting over 100 civilian casualties. Germany had brought the war to England's largest city. [24]

The raid precipitated a crisis within David Lloyd-George's coalition government. The Prime Minister asked General Jan Christiaan Smuts to evaluate Britain's current aviation structure. Completing his assignment in record time, Smuts concluded that aviation technology had rendered geography meaningless. In addition, Smuts observed that aviation had dissolved any discernible distinction between Army and Navy flying units. Accordingly, Smuts recommended that Britain combine Naval and Army into a new entity to be called the Royal Air Force (RAF). Smuts also recommended that Britain form a new cabinet post, a Ministry of Air, responsible for both military and commercial aviation development. The Lloyd George government accepted Smuts' recommendations. On April 1, 1918, Britain's independent aviation force came into being.

As an apprentice to the Allies, U.S. pilots flew British or French aircraft as American factories tooled up for plane and engine production. Brigadier General William Mitchell, Assistant Air chief to General John Pershing, discussed aviation's future with General Hugh Trenchard. Despite the Royal Flying Corps tactical support at Amiens, Trenchard had adopted a new doctrine—a strategic strike against enemy factories. The bomber would bring the war to enemy population centers.

At the time, U.S. naval aviation had been lodged in the most prestigious naval bureau, the Bureau of Navigation. Admiral Bradley Fiske transferred aviation to the new Operations branch. Despite its new residence, aviation's activities remained scattered among the navy's material bureaus. Engineering, for example, was responsible for aircraft radio; Construction and Repair aircraft fuselages; ordnance, guns and ammunitions.[25]

Admiral William Benson, Chief of Naval Operations, was not especially taken by a new intruder into the navy's organizational family. Benson actually downgraded aviation. By the last year of the war the CNO had so distributed aviation activities throughout the Navy's bureau system that one commentator observed that Benson had taken a difficult decision-making process and turned it into an impossible one. In the meantime, air technology experienced quantum increases in engine, speed and tactical capability.[26]

After Britain's introduction of the carrier naval aviation proponents sought to convert a fleet collier into an aircraft carrier. Admiral Benson vetoed the project. To Daniels' credit, the Secretary overruled Benson's objection. Thus was born the U.S.S. Langley, America's first experimental carrier.

In the first year of Wilson's administration, Rear Admiral Bradley Fiske sought permission to hold fleet war maneuvers or games. As part of the exercise, the fleet divided its ships into offensive and defensive units. Although, Daniels set May 1917 as a month for such maneuvers he imposed certain conditions. Fiske's diary recorded the following entry on March 17, 1917;

> Secretary said he does not want to have any war game in May which will include the defeat of the U.S. fleet! So all our plans to make the game educational to the people have failed or will fail!

The secretary, in the meantime, designated war game exercises as the "department strategic problem No. 1."[27]

AMPHIBIOUS OPERATIONS

The most notable amphibious action took place in Asia Minor in the First World War. An ally of Germany and Austria-Hungary, Turkey controlled the Dardanelles, a channel linking the Mediterranean Sea and the Black Sea. Turkish forces prevented the Allies from reinforcing Russian troops reeling under General's von Hindenburg and von Ludendorff's German offensive.

In late 1914, Britain and France began to search for an alternative to the western front stalemate. Winston Churchill, head of the British Admiralty, offered an intriguing proposition. The Royal Navy should force the Dardanelles, enter the Sea of Marmora, penetrate the Black Sea and sail to Russia.

A successful envelopment carried a double return; Britain could deliver munitions to besieged Russian troops; Russia could export wheat to England.

With French assistance, the Royal Navy entered the Dardanelles straits in January 1915. Lining the mountains adjacent to the Dardanelles, Turkish artillery fired on the Royal Navy, whose naval guns replied in kind. Although British minesweepers removed Turkish mines from the channel's center, the sweepers neglected the strait's margins—an oversight that claimed three British warships. The battleship loss exacted a psychological toll on Admiral DeRobeck. Churchill exhorted the fleet to press on—in vain. By March, enough was enough. Admiral DeRobeck retired and the fleet sortied back to the Mediterranean. The British experience appeared to validate Lord Nelson's admonition that "a ship's a fool to fight a fort." Unknown to the British, Turkish troops had exhausted their ammunition.[28]

Turning to its infantry and commonwealth troops, Britain embarked on an amphibious assault against the Gallipoli peninsula. Turkey rushed in reinforcements, prodded by a German military advisor, General Liman von Sanders. The British offensive ground to a halt. In early summer of 1915, the British landed troops on Sulva Beach, a northern peninsula foot hold. Once again, British forces failed to advance beyond the beaches and the fighting degenerated into another stalled front. Though devoid of Passchendaele's mud and rain, Gallipoli took on the traits of siege warfare. By December British and Australian troops evacuated the Gallipoli peninsula and did so without loss of life—a withdrawal that did little to enhance the reputation of amphibious warfare.

President Wilson appointed General John Pershing to command an American Expeditionary Force in Europe. British and French officers, scarred, war weary, hardened by three years of fighting, wanted to employ U.S. troops as replacements to putty up their lines. Pershing insisted that his forces operate as an independent unit assigned to their own sector. In the spring of 1918 Germany launched an offensive toward Paris, employing tactics that would prove a forerunner of the blitzkrieg of World War II. The American 2nd Army Division, commanded by General James Harbord, included the Fourth Marine Brigade. The two armies were separated by an open wheat field. Residing in a wooded area, German troops possessed a formidable defense.

At five o'clock in the afternoon of June 5th General Harbord ordered the 4th Marine Corps to take the Belleau Wood. Crossing an open wheat field, Marines fell by the dozens as Germans plied their machine guns back and forth. By the end of the week, the Marines had achieved their objective. Their casualties exceeded 40%. (General Harbord would later wear marine emblems on his army shirt collar to commemorate the Belleau Wood action.) A logistics snarl erupted at Brest France and General Pershing sent Harbord to sort out the inventory tie-up. Major General John Lejeune replaced Harbord as commander of the Army's 2nd Division. A Marine now led an Army Division.[29]

The Marines' experience in France stood apart from their traditional role as gunner's mates and constabulary duty. Now in Europe, the Marines adopted Army ordnance, infantry tactics, and weapons appropriate to siege and open warfare. Oddly, the Army Air Service refused to accept Marine pilot assistance. Marine aviators did manage to carry out raids against German U-Boat pens, however.

LOGISTICS

Fleet logistics in World War I included naval reserve oil, merchant ships, refueling at sea. Oil fired boilers began to replace coal fired boilers. Petroleum emerged as a critical national defense commodity.

Under the Taft Administration the Navy had requested that an oil reserve be set aside in the event of a national emergency. Richard Ballinger, Secretary of the Interior, designated two California areas, reserve #1 and #2 as a naval petroleum reserve. The Wilson Administration set aside a third oil reserve in Wyoming, reserve #3. Presumably the navy's crude oil would remain safe, stored underground.

Naval reserves, however, resided adjacent private drilling operations, often sharing a common pool of crude. Private wells threatened to erode the Navy's inventory of crude oil, activated by the rule of capture, a common law principle. The rule held that whoever discovered oil first had a right to exploit it. Accordingly private drillers had little incentive to sit on any oil find.

After an extended debate, Congress, in February of 1920, passed legislation permitting the Interior Department to lease oil reserves to private firms who in turn agreed to pay the Navy a Royalty fee. Congress imposed a ceiling on Navy receipts, however. The Department could retain half million dollars of royalty receipts. Any amount above that level had to be turned over to the U.S. Treasury Department.

Although Josephus Daniels had accepted the February legislation, he remained apposed to Interior acting on behalf of the Navy. In the spring of 1920, Daniels suggested that a rider be appended to a Naval Appropriation Bill. The rider embodied two conditions. First, the Secretary of the Navy replaced Interior's Bureau of Mines as petroleum administrator. Second, the rider authorized the Navy Secretary to use, sell, store, exchange "in his discretion." The rider, attached to a large naval expenditure act, was passed by Congress in June, 1920. Daniels had pulled off a legislative coup. Indeed, the Secretary was so confident over the fate of his rider that he set up an oil desk in his office before Congress enacted the bill.[30]

The price of oil rose during the First World War. Daniels, in fact, predicted that a barrel of fuel oil would hit $5.00 a barrel. The Navy was not without an oil alternative. Congress had accorded the Navy authority to commandeer fuel oil, reimbursing the petroleum company at prices below market value.

A second logistic facet included the nation's inventory of merchant ships. Given its global empire, Britain had long operated a large merchant fleet. Even before the U.S. entry into the war, William McAdoo, Wilson's Treasury Secretary, concluded that the U.S. should at least equal, if not exceed, Great Britain's merchant fleet.

In 1916, Congress created the United States Shipping Board (USSB) that, among other things, put the U.S. government into the private shipping business. As part of the nation's mobilization effort, the board created a subsidiary, the Emergency Fleet Corporation. The affiliate embarked on a massive merchant ship construction program during the war.

The chairman of the U.S. Shipping Board began to search for fleet operating cost savings. The board purchased reserve oil from (Salt Creek, Wyoming), swapped crude oil for refined oil, and passed the savings to government vessels competing with commercial shipping firms. The fact that Congress permitted government vessels to compete with commercial shipping firms was viewed as a non-issue. Government ships could serve as a bench mark to check the cost effectiveness of commercial ships, now regulated common carriers.[31]

The Navy was cognizant of another oil dividend - refueling ships at sea. Given their limited operating range, destroyers stood as prime candidates for replenishment underway. During the First World War, a young officer, Chester Nimitz, developed an Atlantic refueling system for destroyers destined for Queenstown, Ireland.

SOURCING

The Navy faced the two choices in securing its material and supplies - obtain assets from its shore establishment or purchase supplies and ships from the nation's industrial sector. In 1914, Congress passed legislation requiring that warships be constructed at government navy yards. The navy could accept bids from private shipyards providing that navy yards were booked to capacity. The next year, congress provided funds to expand navy yard facilities.[32]

Wilson, Daniels and Congress were intent on confining warship construction to the navy's shore establishment. Years later, the Aluminum Corporation of America would be hauled into court on grounds that building ahead of demand violated the nation's antitrust laws.[33]

Congress did more. Lawmakers approved the construction of a naval aircraft factory, a steel mill, a munitions plant, a helium factory, oil reserves, a paint factory, clothing shops. Congress did block Daniels from putting the Navy into petroleum refining activities. On another occasion, Secretary Daniels offered to build merchant ships at government navy yards. That required that navy yard be given exclusive contracts for the construction of fleet oilers.[34]

Daniels was more than willing to share his vision of the Navy Department with U.S. lawmakers. Renamed a Department of Marine, the entity would include the Coast Guard, Panama Canal ships, the Light House Administration and the Army Transportation Service. That was a first step. Daniels wanted U.S. government to own, operate, control and manage the nation's telephone, telegraph, wireless, oil, coal, steel and merchant shipping—all grounded on the premise of national security.[35]

In enlarging the role of the public sector one matter persisted; were government yard employees as efficient as their private yard counterpart? To shed light on the issue, Alfred Meyer, Taft's Secretary, retained an outside accounting firm to audit yard operations. The firm concluded that yards were labor intensive and expensive, a finding that prompted Meyers to phase down and consolidate the Navy's shore establishment.

Daniels, by contrast, assigned a group of Naval officers to evaluate yard building costs. The officers concluded that if the yards reallocated their overhead expenditures, they could achieve a saving of 33%. One salient difference between the two studies turned on the issue of overhead expenses, i.e. the assignment of officers' salaries, yard capital expenditures and electric bills assigned to a particular ship under construction. Here Navy Yards enjoyed considerable accounting discretion.[36]

At the turn of the 20th Century, the U.S. industrial sector experienced a seismic shift. Before, a product remained fixed, as workers moved from one bench to another. As craft machinists controlled the plant's output, automatic machinery began to threaten the status of craft workers. Non-skilled labor now operated specialized machines that punched out a standard component. The next step inverted a plant's output flow. Here the product moved down an assembly line, the employee remained stationary. More important, assembly line workers did not require specialized training associated with trade union practices. The results eroded trade union authority. Under a mass production protocol the craft union employees felt besieged and under assault.

Navy yards became intrigued with the productivity of modern manufacturing practices. In 1911, George Meyer, Secretary of the Navy in the Taft administration, ordered the battleship USS New Hampshire, to take its overhaul work to the navy's Norfolk yard, Virginia. Meyer directed Norfolk's civilian employees to fill out index cards indicating the work performed and the time consumed in completing a particular task. Norfolk's employees balked at the request. In fact, some 400 workers walked out of the plant. Meyer directed the New Hampshire to proceed to the navy's Brooklyn yard. Presumably the New York employees were more amenable to the secretary's card request.[37]

The standoff between civilian employees and the Navy Secretary erupted over the introduction of a production technique expounded by a machinist turned foreman, Frederick Winslow Taylor. Born and raised in Philadelphia, Taylor came from an affluent family, and was sent to Phillips Exeter Academy before applying to Harvard University.

Taylor enrolled in a geometry and algebra class taught by George A. Wentworth who employed an idiosyncratic learning routine. He assigned a mathematics problem each day and allotted the student's time to work out the answer. Once completed, a student raised his hand. It took some effort for students to adjust to Wentworth's classroom regime. With practice, however, students were able to reduce their computation time, measured by a stop watch lying on Wentworth's desk. Frederick Taylor not only absorbed his Exeter lessons, he applied the technique when he became a foreman at the Midvale Steel Company outside of Philadelphia.[38]

An astute observer, Taylor concluded that workers spent half their walking—searching for tools, adjusting machines, rummaging for material. Taylor reasoned that if a worker allotted more time at a machine, employee productivity would improve. Taylor proceeded to deconstruct each work assignment into a series of steps. Measured by a stopwatch, he cut out wasted movements. Taylor then tied an individual's wage compensation to his output. Committed to production flow, Taylor rearranged the plant's floor layout, introduced quality control, instituted inventory management and adopted cost accounting. Taylor even designed equipment to reduce worker fatigue.

Taylor was essentially saying that plant foremen, using guess work and past practices, simply did not know how to manage their workers. The application of science to the production process would at least give a manager knowledge, data, and information. Taylor would call his technique, Scientific Management.[39]

After leaving Midvale Steel Company, Taylor began to proselytize his production methods to firms and companies in the US economy. Companies that adopted Taylor's practices recorded impressive results; productivity soared, unit costs fell, revenues expanded, paving the way an increase in employee wages. As a bonus, customer prices declined. According to Taylor, all parties to the production process gained from the adoption of scientific management. Production, in short, was no longer a zero sum game.

The Ford Motor Company exemplified Taylor's shop floor doctrine. Ford's assembly line enabled the company to cut the price of its model T from $850 in 1908 to $360 by 1916. Nor was Ford not alone. International

Harvester, Western Electric, Westinghouse, General Motors, du Pont adopted elements of Taylor's shop floor practices as well. [40]

The navy's line officers were obviously taken by Taylor's shop protocol. Despite the navy's success in its recent war with Spain, U.S. officers remained troubled by their gunnery record. U.S. targets hits averaged three and half percent. After the fleet adopted Taylor's time and motion studies, target accuracy rose to 33%. Line officers now became Taylor converts. [41]

The Navy's shore establishment was equally intrigued by scientific management principles. Yard administrators introduced time studies at the navy's Mare Island Yard on the west Coast and the Norfolk Yard on the east coast. Yard productivity rose. The U.S. War Department also adopted time and motion studies and bonus incentives at its Springfield and Watertown arsenals in Massachusetts. Labor began to view Taylorism with utmost suspicion and took to the pavement. [42]

The application of science to the shop floor migrated to Europe and the Far East. Vickers in the UK; Renault and Michelin in France; Siemens and Bosch in Germany; Mitsubishi and Nippon Electric in Japan. All adopted varying application of time studies and piece work incentives.

Japanese academicians translated Taylor's scientific management text and the book sold over a million copies. Following a plant visit to the U.S., Godo-Takuo, a naval constructor at Japan's Kure Navy Arsenal, introduced "...division of labor, centralized planning, stopwatch time study, cost accounting, Gantt's chart tracking and instruction card procedures." The arsenal would later design and manufacture torpedoes. At the same time, U.S. yard unions were successful in blocking "the practice of running two or more machine shop at the Newport torpedo station."[43]

Samuel Gompers, President of the American Federal of Labor AFL, was concerned that plant automation eroded a machinist's skill, status and compensation. He was joined by James O'Connell, a machinist's union president, who felt that once government arsenals adopted Taylor practices, private corporations would be encouraged to embrace the management model as well. That concern, however, was reserved for private consumption. For public consumption, labor leaders insisted that time and motion studies favored younger employees at the cost of older workers. Gompers and O'Connell put their case in term of worker's health and industrial democracy.

By December 7, 1914, Gompers decided to "...do our best as rapidly as we can to remove its scientific management evil and degrading influences." The AF&L turned to Congress for help. In response, a House of Representatives Labor Committee held hearings on Taylor's stop watch methodology. Union representatives testified that scientific management was nothing but a production line speed up. Frederick Taylor recited the productivity gains derived from shop floor reform. The Army's chief of ordnance supported Taylor.[44]

Two years later in 1915, the House and Senate legislated a ban on time studies and incentive wages at government plants. The paragraph took the following form;

> That no part of the appropriations made in this Act shall be available for the salary or pay of any officer, manager, superintendent, foreman, or other person having charge of the work of any employee of the United States Government while making or causing to be made with a stop watch or other time-measure device a time study of any job of any such employee between the starting and completion thereof, or of the movements of any such employee while engaged upon such work; nor shall any part of the appropriations made in this Act be available to pay any premium or bonus or cash reward to any employee in addition to his regular wages, except for suggestions resulting in improvements or economy in the operation of any Government plant.[45]

Had Congress accepted Daniels' vision of government ownership of telephones, telegraph, wireless, coal, oil, steel, merchant vessels and aircraft production, the stop watch ban would migrate to additional government arsenals. In any event congress banned existing arsenals, yards and stations from adopting scientific management techniques. Civilian trade unions had scored a significant victory.

By this time Congress had blanketed yard employees' with civil service protection. Newport's torpedo managers, for example, could not dismiss an employee alleged to be "soldiering" until after a hearing and a lengthy appeals process. Any plant manager who had the temerity to admonish an employee for unsatisfactory work could be expected to hear from a member of Congress. During congressional elections, local yard employees transformed themselves into a political machine on behalf of a resident incumbent. Navy yard employment was now politicized.[46]

As Navy Secretary, Josephus Daniels enthusiastically supported the Taylor ban. Although Franklin D. Roosevelt, Assistant Secretary, complained that one yard job estimated to be completed in nineteen hours actually consumed eleven days, he nevertheless sent yard promotion lists to the head of New York's Democratic Party for clearance and approval.[47]

Congress's ban on modern management practices transformed Navy Yards into an insulated, cloistered environment. Over time, civilian yard employees' wages and benefits crept higher than private ship yards workers. Nor did the distinction end there. Commercial yards were required to submit firm prices for government work; the navy yards required to post estimated costs only. The outcome was predictable. Government yards generally underbid private yards. After the contract was awarded in-house, Secretary Daniels admonished private yards to "sharpen their pencils."[48]

The rules for tardy vessel delivery were strikingly different as well. If a private yard missed a delivery deadline, the firm suffered a financial penalty. To protect themselves, commercial shipbuilders took out an insurance bond. Government yards incurred no such penalty. Cynics suggested that Navy Yard employees rallied to the cry, "don't give up the ship."[49]

Ensconced in each yard, the material bureaus staffed, controlled and administered their respective work assignment. The Bureau of Construction and Repair was responsible for a ship hull and design; the Bureau of Steam Engineering, ship machinery, power equipment, electrical and wireless apparatus; the Bureau of Ordnance, ammunition and guns; the Bureau of Yards and Docks, power plants and associated equipment. The absence of any overall administrative coordination occasionally resulted in completed ships with unprotected ammunition hoists and/or limited operating range.

Established government yards, New York, Philadelphia, and Norfolk, handled heavy warship construction; Boston and Charleston, smaller yards concentrated on destroyers, destroyer escorts, ship parts and various supply apparatus. The Boston yard specialized in chains; Mare Island yard, ship paint.

Daniels was convinced that U.S. corporations were prone to fix prices in order to fund dividends to their affluent shareholders. In contract negotiations, Navy buying officers had to be ever vigilant against the possibility of market collusion. No such conduct attended government employees who equated their compensation and benefits with the public interest.

Daniels obviously became exercised when the Navy received identical bids from the nation's U.S. steel companies. In response, Congress put the Navy into the steel business. One steel executive asked Daniels how he intended to operate the armor plant. The secretary replied that he would commandeer the executive, make him a captain and order him to run the plant.[50]

Congress backed Daniels's drive to permit Navy yards to make components in house. On occasion the U.S. Senate challenged that policy. Senator Warren Harding (Ohio) questioned the navy's ability to diversify into steel business and cited the navy's own yearbook. On a cost per ton basis: Krupp posted $490; Austria $511, Italy $405, Germany $490, France $460, England $503, Russia $368, Japan $490. The U.S. posted $425 per ton. Senator Harding reminded Congress that it funded government over-runs.[51]

The Senate dismissed Senator Harding's cost data out of hand. Some colleagues argued that a concentrated steel industry preferred to engage in non-price as apposed to price competition. That theoretical nicety was beside the point. Senator Benjamin Tillman of South Carolina, joined by Daniels, had long sought to put the Navy in the armor plate business. Instead of erecting a plant in Charleston, South Carolina, the Wilson government placed

an armor factory in Charleston, West Virginia. Congressional seniority once
again emerged as the coin of the realm.

MOBILIZATION

Prior to April 1917 the U.S. economy was essentially engaged in civilian
output and production. After April, the armed forces demanded war material
ranging from ships to munitions, from horse fodder to Eagle Boats. It was not
long before the nation's need for war material overwhelmed the capacity of
government arsenals. The Wilson administration turned to the private sector
to balance the nation's restricted output on one side against an insatiable
material demand on the other. Prices began to rise.

The Wilson government was not without an industrial mobilization option, however. For two years, a former president of the Chicago Diamond
Match Box Company, Edward Stettinius, had acted as a purchasing agent on
behalf of the British government. Stettinius, joined by two dozen staff members, operated out of the New York office of the J.P. Morgan. Stettinius
generally arrived at the office at nine in the morning and worked until midnight. In adhering to the bosses' regimen, his associates labeled themselves,
"SOS", slaves of Stettinius.[52]

All told, the Stettinius group purchased some $3 billion dollars' worth of
forage, horses, barbed wire, cotton, gun power, picric acid, steel—the list
was all encompassing. Nor did the Stettinius group confine themselves to
military hardware. They secured capital funds for U.S. plant expansion; set
up truck, rail, and ship delivery schedules; unraveled a railroad tie up on the
U.S. east coast. So pervasive was Stettinius's buying operation that no less
than General Eric Ludendorff asserted that Edward Stettinius was worth an
army corps to the allies.[53]

The breadth and reach of Stettinius's purchases were of so impressive that
Lloyd-George, Britain's coalition Prime Minister suggested that Stettinius
spread his contracts evenly between Democratic and Republican companies.
Stettinius replied that he purchased the best product at the best price—a
practice that subordinated political affiliation to product merit.

Edward Stettinius offered his services to the Wilson Administration. Here
was an individual whose talent, experience and industrial knowhow, backed
by a seasoned staff, could make a singular contribution to the nation's war
effort. President Wilson rejected Stettinius' offer out of hand. The former
match box executive was a J.P. Morgan partner. Wilson regarded J.P. Morgan as a Republican bank. In any event, the President did not want to alienate
a key political ally, Samuel Gompers, President of the American Federation
of Labor.[54]

Woodrow Wilson thus elected to embark on his own mobilization plan. He set up a National Defense Advisory Committee, composed of his cabinet officers. The advisory committee set output quotas, ordered material production, logged countless meetings, issued a series of directives. The heroic effort yielded little traction. Some observers blamed the mobilization problem on bad luck, the severe winter of 1917-1918. Others thought that the President's cabinet was over-burdened by their dual administrative assignment. Professor Albert A. Blum put the matter directly. Industrial mobilization, he said, was "...substantially stalled by the logjam in the White House."[55]

By 1918, the President turned to a Wall Street investor and Democratic Party contributor, Bernard Baruch. Wilson appointed Baruch as head of the War Industry Board. Baruch first requested that the nation's antitrust laws be suspended for the duration. Next Baruch issued broad industry production quotas. Finally, he turned to industry trade associations, directing them to allocate war output among and between industry members. Under Baruch's leadership, U.S. war production began to take hold.

The unexpected then occurred. Germany agreed to an Armistice and the guns on the western front fell silent. By November, 1918 American factories had delivered 143 pieces of artillery, six large tanks, a dozen airplanes. The world war had ended before U.S. production could make its mark. Industrial mobilization was not an instantaneous process.

Government expenditures predictably soared during the war. Congress imposed a series of taxes to fund the nation's military effort; an income surtax, a tax on corporate savings, a tax on the nation's munitions industry, a doubling of personal income taxes, a luxury tax, and an estate tax. Tax receipts, however, failed to keep pace with government expenditures and the nation's debt jumped from $1.1 billion in 1914 to $24.3 billion by 1920. Government employment rose from 500,000 in 1914 to nearly 700,000 by 1918. Personnel assigned to the Internal Revenue quadrupled from 4,000 in 1913 to 18,000 by 1920. Historians are in general agreement that the First World War was funded largely by taxes on "excess" corporate profits.[56]

INTER-GOVERNMENT COORDINATION

The nation's armed services attempted to make contingency plans prior to the August crisis of 1914. The President, equating preparation with belligerence, banned the Army-Navy Joint Board from meeting without his approval. Presumably, a statement from the U.S. State Department as to U.S. foreign goals might have enabled the Army and Navy to tailor and adjust their force requirements. But such intergovernmental coordination proved at best episodic. Much to the frustration of Secretary Robert Lansing, Woodrow Wilson preferred to be his own Secretary of State.

Post Armistice Developments, 1918-1921

Following the Armistice, the Wilson administration addressed several inter-related issues; military preparations, government control in a post war economy, government fiscal restraints, Navy department restructuring and surprisingly, the eruption of a U.S.-Japanese war fever.

Preparation

In less than four weeks after assuming his second term in office, Woodrow Wilson asked Congress to declare war against Germany and Europe central powers. Having invaded Mexico, the U.S. Army was distracted in preparing for any European venture. Though relatively small compared to its allies, U.S. casualties on western front were not insignificant. The army originally placed its casualties around 100,000. Thomas Fleming, the historian, noted that if one included deaths immediately after the war, the figure rose to 460,000. After the conflict, Congress turned to the question of military preparedness. Admiral William Sims, returning from Europe, was unusually harsh. Sims stated that, under Daniels, the navy was ill-prepared, "…in spite of the fact that war had been a probability for at least two years and was in fact, imminent for many months before its declaration."[57]

Sims essentially charged Daniels with administrative incompetence. Daniels rejoined that not only was the Navy prepared for the European war, but that the Navy had won the war. Daniels then fired his own ad hominem. Admiral Sims, he said, was posturing in order to solicit an award from the British government.

The Sims/Daniels quarrel spilled into the Navy's officer corps. Operation's staff officers cited their work in anticipation of fleet readiness. Line officers, on the other hand, tended to side with Sims. After several days of hearing, all parties appeared exhausted, if not slightly embarrassed. The Senate issued its report. A majority committee report sided with Sims; a minority with Daniels.[58]

In the early months of World War I, Congress authorized the production of battleships and battle cruisers, embodied in the naval appropriations act of 1916. As noted, the effectiveness of German U-Boats prompted the Navy to shift from heavy warships to anti-submarine vessels, destroyers and destroyer escorts. Having deferred battleship construction until the German submarine crisis eased, Wilson, through Daniels, announced a resumption of dreadnought construction. The decision sent off alarm bells in Britain and Japan.

Privatization

At the turn of the century, the Navy vested an interest in wireless communications. In 1912, congress legislated a wireless act that accorded the Navy exclusively control over long wave frequency operation. Amateur wireless operators were relegated to the "useless" segment of the frequency spectrum—200 meter/second or smaller. The act also authorized the President to seize amateur wireless operations in the event of war. Prior to the U.S. entry in to the European war, wireless patent ownership was so scattered that manufacturing a wireless set was tantamount to inviting a patent infringement suit.

Daniels encouraged the Navy to purchase domestic wireless stations owned by foreign countries, notably Germany and Britain stations. The secretary's strategy was clear. The nation's wireless operations constituted a natural monopoly, and the Navy would be that monopoly. Senator Joshua Alexander filed legislation to legitimize the navy's wireless takeover.

The Senate Merchant Marine and Fisheries Committee held hearings on the navy's actions. Representatives from American Marconi, a British affiliate, testified against the legislation. Navy officers supported the legislation. When questioned why they approved the navy's acquisition of American Marconi, one officer replied the Navy was doing so at the requests by U.S. Shipping Board. [59]

Committee members were so critical of the Alexander legislation, that the bill never reached the floor of the Senate. The Navy abandoned its attempt to be the chosen instrument of U.S. commercial wireless operations. Still, the Navy had a fallback position. If the U.S. disapproved of a government monopoly, perhaps a private monopoly would suffice.

It all started when G.E. informed the Navy of its intent to sell its transmitting alternator to British Marconi. The Navy faced the prospect that American ship wireless equipment would be dependent on a foreign power. The Navy asked G. E. to retain its wireless investment within the U.S. In response, G.E. formed a patent holding company, permitting owners to manufacture and supply wireless equipment free from litigation. The holding company, the Radio Corporation of America (RCA), would be owned by G.E., Westinghouse, and AT&T, the telephone company. The firms proceeded to divide and allocate the radio market— RCA would sell vacuum tubes and point to point maritime services; G.E. and Westinghouse would allocate the wireless set market on a 60:40 basis; and AT&T would engage in wireless telephone and transmission equipment. As a bonus to the Navy, an admiral would sit on RCA's Board of Directors. The Wilson Administration had converted a wireless gridlock into an industry with a promising future. [60]

Wireless amateurs operating their own segment, employing crystal sets, then upgraded to vacuum tubes. Prior to the war, some 4000 amateurs had

formed the American Radio Relay League publishing *QST*, a monthly magazine. Confined to 200 meters amateurs envisioned a national network of
wireless linkages.

The First World War interrupted the league's vision of national network.
President Wilson instructed Daniels to shut down wireless sets and Daniels
ordered amateurs to dismantle their equipment. When the war ended, however, wireless amateurs became concerned by the Navy's role as a wireless
monopoly. The AARL apposed Daniel's takeover of commercial wireless
operations.[61]

U.S. amateurs were wary of the U.S. Post Office as well. To support its
air mail service, the Post Office planned to build wireless stations throughout
the country. The Post Office requested a ban against amateurs operating
within one mile of its transmitters.

Although radio amateurs felt besieged by two government agencies, their
concerns began to ease by 1920. Congress ordered the Navy to sell its wireless acquisitions. The Post Office returned operation of the Bell Telephone
Company back to its owners. The government lifted its ban on amateur
wireless operations. Daniels' monopoly vision had experienced a setback.[62]

Government Budget

The First World War dramatically altered the nation's public finances. Prior
to the war, the U.S. federal debt stood at roughly one billion dollars; by 1920,
U.S. debt topped 24 billion dollars. Despite the imposition of wartime taxes
the nation's expenditures outpaced its tax revenues, the difference funded by
bond sales. In addressing the nation's deficit and debt, Congress passed a
national budget act. President Wilson vetoed the measure on grounds that the
legislation denied him authority to dismiss the Comptroller General.

The House of Representatives corrected the President's concern. A new
bill passed House and was sent to the Senate. The President's party engaged
in a filibuster and killed the budget bill. Though stalled, the concept of a
federal accounting office would remain very much alive.[63]

Departmental Restructuring

A third policy issue focused on the Navy's administrative structure. During
the war, Congress appropriated a billion dollars to fund the department's
shore establishment, bureaus and fleet. Naval spending was administered by
Rear Admiral Samuel McGowan, Chief of the Bureau of supplies and Account. In letting contracts, McGowan resorted to competitive bids. No scandal attended the navy under McGowan's watch.

Franklin D. Roosevelt's nearly eight years as assistant Navy secretary had
left him with decided views on the department's operation. At the outset

Roosevelt asked Congress to give the Navy secretary authority to allocate expenditures among and between the Navy's bureau chiefs. Roosevelt argued that the bureau chiefs were not amenable to cooperation, comparing them to "dogs in the manger." Moreover, the assistant secretary claimed that 2,000 out of 10,000 Navy Yard employees were unproductive and redundant. Finally, Roosevelt questioned the Navy's promotion system driven by seniority rather than by personal merit. [64]

Franklin Roosevelt added that officers educated at the U.S. Naval Academy were ill-trained to administer shipyard activities. Vessel construction constituted a discipline unto itself. Nor was the touted efficiency of Navy Yards particularly persuasive. Roosevelt insisted that the Charlestown Navy Yard had taken eleven days to complete a job when it could have been done in nineteen hours. Roosevelt speculated that if a private business ran itself like the Navy, the firm would soon be insolvent. Finally, Roosevelt insisted that oppressive federal regulation of private firms was the surest way of throttling a company's creative and productive energy. The Assistant Secretary then uttered the unthinkable. Congress, he said, was responsible for the department's difficulties. [65]

Senator Frederick Hale, chairman of a subcommittee on Naval Affairs, asked Franklin Roosevelt to share his insight, experience, and recommendations with the committee. Roosevelt responded: "I am absolutely apposed to any action by your subcommittee looking to changes in the existing organization... I don't believe the time has come for a careful examination of the broad subject. We are too close to the war to understand its lessons."

The senate took notice and the Navy's organizational structure emerged unchanged. [66]

War Fever-Japan

In the fall of 1918 President Wilson turned his attention to peace negotiations with Britain, France, Italy and Japan. The President proposed a fourteen point plan, the centerpiece calling for the creation of an international body, a League of Nation. The league would forestall international disputes before they erupted into open warfare. Presumably Wilson's fourteen points influenced Germany to accept a November armistice in the fall of 1918.

Whether measured as casualties, capital destruction or government finances, the Great War had proved a traumatic experience for France and Britain. Both nations were now billions of dollars in debt to the U.S. After the war, U.S. allies sought compensation. Wilson wanted no part of the spoils of war. Rather, the president placed a premium on selling his league concept of fourteen points. Although Lloyd George suggested that Moses had confined his commandments to ten, Britain agreed to support Wilson's League of Nations.

While itemizing an accounting bill to be presented to the German government, France was receptive to Wilson's league concept. Among other considerations, the bill included the cost of casualties, destroyed property and pensions for surviving French veterans.

Japan had undertaken convoy duty for British common wealth troops on route to the Mideast. In return, Japan and Britain quietly agreed to divide German Pacific holdings. Britain would take islands south of the equator; Japan would seize the Micronesian Islands, north of the equator.

Italy could not resist compensation as well. Victorrio Orlando, the country's Prime Minister, planned to annex Fiume, an Austrian-Hungarian Port on the Adriatic. When Orlando resisted Wilson's plea to forgo the booty of war, Wilson, via an Italian newspaper article, took his case to the Italian public. Upon reading Wilson's article, Orlando stormed out of the Paris meetings, taking with him a League vote. With exquisite timing, Japan's representative insisted that they retain control of Germany's Pacific islands, residing north of the Equator. Unless Japan's request was honored, Japan threatened to walk out of the peace talks placing Wilson's League of Nations at risk. [67]

Wilson's management style now came into bold relief. Although Robert Lansing, secretary of state, had accompanied the President to Paris, Lansing essentially remained in his hotel room, writing memos to the President. Wilson dismissed the memos on grounds that "lawyering" would clutter up the peace process. [68]

State Department advisors, on the other hand, did suggest that the allies return the islands to Germany as a first step. State recommended that the U.S. take control of the islands in lieu of German reparations. Wilson rejected that proposal out of hand. Wilson's naval advisors, by contrast, concluded that the islands were incidental to U.S. national interest and that Japan could retain control of the Marshalls, Caroline and Marianas provided they remain unfortified. One naval historian termed the Navy's position as an act of "high statesmanship." [69]

The President could have invited a senior member of the Republican Party as a bipartisan effort in foreign policy. Instead Wilson selected a relatively unknown member of the opposition party. Upon his return to the U.S., the Senate began debating the content of the Versailles Treaty. Critics of a League concept insisted that its membership compromised U.S. foreign policy independence. Wilson argued that the league was intended to defuse the onset of another world war. Negotiations continued between the Senate and President in the spring of 1919. The Senate did fashion a compromise. The President, however, instructed his party to veto the bill and they followed his wishes. [70]

President Wilson now elected to take his case to the American public. He began a series of whistle stop speeches across the country. In Colorado,

Wilson suffered a massive stroke and collapsed. He would be essentially bed ridden during his remaining months in office.

As noted, the president had announced his intention to double the size of the navy 1916 program, after the November Armistice. Britain proceeded to embark on a new warship – the Hood class. Japan announced a construction program of eight battleships, eight battle cruisers in eight years. That Japan viewed the U.S. construction program with some alarm was seen by General Kojiro Sato statement in the Peking Daily News,

> when America's program of naval extension is completed she will have 40 old and new battle-ships, 37 cruisers, 258 torpedoes – destroyers, more than 300 submarines, and 5,000 seaplanes. There will be a corresponding increase in mercantile shipping, whish on completion of the plans underway, will enable America to put in conversion 1,039 ships of various descriptions, totaling 5,924,200 tons. America has on army of 120,000 men in prewar days by is contemplated to increase the forces to 300,000 at a bound. [71]

Despite an anglophobic tradition among some U.S. officers, the Navy began to detect unsettling developments in the Pacific. Japan's occupation of the Micronesian Islands intersected a communication line between Hawaii and the Philippines. More worrisome, Japan invited a British aviation mission to advise its navy on carrier aviation. Japan then ratcheted up its naval budget from $85 million in 1918 to $250 million by 1921. Japan received seven German submarines as war reparations. Japan then proceeded to import some 200 German diesel experts as engine consultants. [72]

In the summer of 1919, Daniels transferred half of the U.S. fleet to the Pacific. In December 18, of that year the Joint Army-Navy Board designated Pearl Harbor as the apex of a defense triangle that included Puget Sound in the north, and the Canal Zone in the south. [73]

Admiral Robert Coontz, succeeding Admiral William Benson as the Navy's new CNO, ordered the U.S. Marine Corps to organize an east and west coast Expeditionary Force of 6,800 troops each. The force, said Coontz, should be prepared to move into action within 48 hours. If the occasion arose, the west coast contingent was designated to seize key islands in the western Pacific, the Marshalls and Carolines. An east coast force was assigned to occupy Haiti, Santo Domingo and Cuba. The Marines were beginning to inch away from their traditional constabulary duty. [74]

Intelligence reports did little to ease the U.S. Navy's concern over Japan's intentions. In March of 1920, the Yakumo, a Japanese warship, visited Pearl Harbor for fuel and water. U.S. Army intelligence officers caught Japanese officers, in civilian attire, measuring the dimensions of Pearl Harbor near Diamond Head. Japanese officers possessed a telescope and a range finder. Although the Yakumo departed the next day, the incident reinforced the navy's concern over Pearl Harbor's vulnerability. [75]

That same month, the Navy decided to dredge Pearl Harbor's S channel to accommodate additional battleships. The Navy's Hawaiian defense plan included facilities for torpedoes, power supplies and fuel oil storage. In testimony before the House Naval Affairs and the Senate Naval Affairs Committee, Secretary Daniels and Admiral Coontz stated that the U.S. should bulk up its defense on the west coast, Pearl Harbor and the Philippines. Daniels asked the committee to go into executive session. *The New York Times*, *The Washington Post*, and *The Army-Navy Journal* reported that the Secretary had described the Japanese Fleet as a "menace." Daniels denied the description and said that relations between the U.S. and its former "associate" were most cordial. [76]

In 1920, the navy's General Board designated Pearl Harbor as a "great refueling station." Storing oil in floating barges, said the General Board, would be insufficient. To expedite ship refueling, the Navy required a pumping station and a pipeline to Ford Island. In the words of the General Board's, "when arrangements have been made for rapid refueling of these vessels the most important part of the development of Pearl Harbor will have been taken care of." [77]

By the summer of 1920, the Chief of Naval Operations ordered the Marines to Guam. He requested a list of ordnance, gun placement and aviation requirements. Daniels directed the Marines to "take all means necessary to execute plan." The Marines, by this time, had renamed their War Plans Section the Division of Plans and Training. [78]

The U.S. Navy apparently learned that Japanese merchant vessels had been assigned specific locations throughout the Pacific Ocean. To a naval professional, merchant hulls, as auxiliaries, were less than benign. Fleet auxiliaries could supply a naval offensive. The intelligence information provoked still another question; was the Japanese Navy contemplating a move in the near future? No one knew, of course, Admiral Albert Gleaves did report that a Japanese naval official boasted that his country's troops could take the Philippines within 48 hours. Gleaves added that Japan might strike "without warning." A U.S. Attaché memo reported that 8,000 Japanese resided in the Philippines and that Japan regarded the U.S. control of the Philippines as a "… fist in the face of Japan." [79]

In the last months of his eight year term in office, Secretary Daniels found little time to relax. In January 1921, the navy issued an "are you ready?" memo to the fleet. Commanders were instructed to respond to some 18 questions including armor, propeller shaft, ship control, fire control, sea worthiness, steaming radius. Concerned over Japan's submarine development, the War Plans Division of Operations speculated that it was not inconceivable that Japanese submarines could isolate Pearl Harbor from the U.S. mainland. Absent any oil facilities on the island, the fleet at Pearl Harbor might be subject to a Japanese oil embargo. [80]

In January 1921, Secretary Daniels wrote the House Majority Leader requesting an extension of the President's oil commandeering authority until June 30, 1922, an authority that would enable the Navy to purchase oil at rates 25% below market prices.

On February 28, 1921, Secretary Daniels signed off on a Pearl Harbor defense plan that contemplated a mine barrage at the harbor's entrance, personnel passes for base access, the construction of a base fence, a "portable wireless truck," operated by Marines to protect water and electric utilities facilities. The plan included an oil storage facility on the island as well. [81]

Daniels proceeded to solicit leasing bids for oil reserve #1, Elk Hills. The lease contemplated that a petroleum company reimburse the Navy in the form of fuel oil. Under the legislation of June 4, 1920, congress empowered the navy secretary to use, store, and sell exchange reserve oil "in his discretion."

In two short years after the guns had fallen silent in Europe, Secretary Daniels transferred half the fleet to the west coast, approved a Pearl Harbor defense plan, and proposed oil storage facilities on both U.S. coasts as well as Hawaii. Daniels, sought lease bids on the Elk Hills naval reserve in California. The navy's War Plan Section estimated the requirements associated with a trans-Pacific offensive. The west coast fleet would sail to Hawaii, refuel, and move to Eniwetok as an advanced base. From Eniwetok, the U.S. fleet contemplated a sortie toward Truk in the Carolines as a prelude to decisive action in the Philippines.

The country was now about to experience a political transition. Warren Harding would assume the presidency on March 4, 1921. Having served on the Senate's Naval Affairs Committee, Harding was known as a "big navy" man.

NOTES

1. Michael A. West, "Laying the Foundation: the Naval Affairs Committee and the Construction of the Treaty navy, 1926-1934," Ph.D. dissertation, Ohio State University, 1980; see also Thomas Fleming, *The Illusion of Victory: America in World War I*, New York: Basic, 2003), 379.

2. Fleming, 311; also Paolo Coletta, *Admiral Bradley A. Fiske and the American Navy*, (Lawrence: The Regents Press of Kansas, 1979), 232.

3. Coletta, 106; also Melvin I. Yrofsky, "Josephus Daniels and the Armor Trust," *North Carolina Historical Review*, "...entered the department with a. profound suspicion that whatever an Admiral told him was wrong..." p. 240.

4. Bradley A. Fiske: *From Midshipman to Rear Admiral*, (London: Werner Laurie, 1930), 604-608.

5. Robert H. Connery, the navy and the Industrial Mobilization in World War II (Princeton: Princeton University Press, 1951), 20.

6. Robert Greenhalgh Albion, *Makers of Naval Policy, 1798-1947*, (Annapolis: Naval Institute Press, 1980), 41.

7. Robert R. Beers, "The Development of the Office of the Chief of naval Operations," 10, *Military Affairs*, (Spring, 1946), 62-64.

8. Gerald E. Wheeler, Admiral William Veazle Pratt, "A Sailor's Life," (Washington: Department of the Navy, 1974). Also J.A. Stockfisch, *Plowshares into Swords, Managing the American Defense Establishment*, (New York: Mason and Lipscomb, 1973), 55-58.

9. Carroll Kilpatrick, *Roosevelt and Daniels: A Friendship in Politics*, (Chapel Hill: University of North Carolina Press, 1952), 68.

10. Frank Freidel, *Franklin D. Roosevelt: The Apprenticeship*, (Boston: Little Brown, 1952), 239; also William J. Williams, "Josephus Daniels and the U.S. Navy's Shipbuilding program during World War I," 60, 1, *Journal of Military History*, (January, 1996): 22.

11. E. David Cronin, ed., *The Cabinet Diaries of Josephus Daniels, 1913-1921*, (Lincoln: University of Nebraska Press, 1963), 68.

12. Ronald Spector, *Admiral of the New Empire: The Life and Career of George Deney*, (Columbia: University of South Carolina Press, 1974), 199.

13. James C. Bradford, ed., *Admirals of the New Steel Navy: Makes of the American Naval Traditions, 1880-1930*, Annapolis: Naval Institute Press, 1990), Benjamin Franklin Cooling, "Bradley Allen Fiske: Inventor and Reformer in Uniform," 134.

14. Naval Administration, *Selected Documents on navy Department Organization*, 1915-1940, Naval War College Library, July 7, 1945, 3.

15. Elting E. Morison, *Admiral Sims and the Modern American Navy*, (Boston: Houghton Mifflin, 1942), 368-369.

16. Lorrelli Barnett, *The Swordbearers: Supreme Command in the First World War*, (New York: William Morrow, 1964), 176.

17. John B. Hattendorf, B. Mitchell Simpson III, John R. Wadleigh, *Sailors and Scholars: The Centennial History of the U.S. Naval College*, (Newport: Naval War College Press), 142.

18. Joseph L. Morrison, *Josephus Daniels: The Small d Democrat*, (Chapter Hill: University of North Carolina Press, 1996), 68; see also Freming, 192.

19. Raymond C. Gamble, "Decline of the Dreadnought: Britain and the Washington Naval Conference, 1921-1922," Ph.D. dissertation, (University of Massachusetts, 1993), 130.

20. Hugh Popham, *Into Wind: A History of British Flying*, (London: Hamish Hamilton, 1969), 59.

21. Caroll Kilpatrick, *Roosevelt and Daniels: A Friendship in Politics*, (Chapter Hill: University of North Carolina Press, 1952), 43.

22. Robert L. O'Connell, *Sacred Vessels: The Cult of the Battleship and the Rise of the U.S. Navy*, (New York: Oxford University Press, 1991), 234.

23. George T. Davis, A Navy Second to None: The Development of Modern American Naval Policy, (Westport: Greenwood Press, 1940), 268.

24. Neville Jones, *The Beginning of Strategic Air Power: A History of the British Bomber Force, 1923-1939*, (London: Frankcass, 1987), 13.

25. Prescott Palmer, "World War I Expansion," Randolph Wiking, ed., *Engineering and American Seapower*, (Baltimore: Nautical and Aviation Publishing Company of America, 1989), 105.

26. Administrative History, U.S. Naval Administration in World War II, DCNO (Air), No. 33, Aviation Training 1911-1939. Navy Department Library, Washington, D.C. 61. Statement by Commander Jerome C. Hunsaker, 47; also Robert Greenlaugh Albion, *Makers of Naval Policy: 1798-1947*, (Annapolis: Naval Institute Press, 1980), 374.

27. Bradley A. Fiske, *From Midshipman to Rear Admiral*, (London: T. Werner Laurie, 1930), 578.

28. Edward S. Miller, *War Plan Orange: The U.S. Strategy to Defeat Japan, 1897-1945*, (Annapolis: Naval Institute Press, 1991), 37.

29. George B. Clark, *Their Time in Hell: The 4th Marine Brigade at Belleau Wood, June 1918*, (Pike: The Brass Hat, 1996), 144.

30. Edwin Denby, Edwin Denby papers, Burton Historical Collection. Detroit Public Library. Letter from Josephus Daniels to Senator Carroll S. Page, Chair, Senator Naval Affairs, U.S. Senate, Box 5, April, 1920.

31. U.S. Senate, Congressional Record, 15, 1, August 11, 1916, 12462. Senator Warren Harding warned that the U.S. Shipping Board competing with U.S. merchant ships would open up a "...veritable Pandoras' Box..." or Regulation.

32. Committee on the Merchant Marine and Fisheries, *The Use and Disposition of Ships and Shipyards at the End of World War II*, United States Navy Department and United States Maritime Commission, Graduate School of Business, Harvard University, June 1945, 202 [cited as Harvard Study].

33. *Harvard Study*, 202, *United States V. Alcoa*, 148, F2d (2d Cir. 1945)

34. Secretary of the Navy, *Annual Report*, 1915, 37; also *American Federationist*, April, 1915, 292.

35. *New York Times*, December 29, 1920, 20.

36. George von Meyer, "Are Naval Expenditures Wasted?" *North American Review*, (February, 1915): 249. Also, Annual Report, Secretary of the Navy, 1914, 15; Oliver E. Allen, This Great Mental Revolution, *Audacity*, 4, 4 (Summer, 1946): 61.

37. Charles D. Wrege, Ronald G. Greenwood, *Frederick W. Taylor: The Father of Scientific Management, Myth and Reality*, (Homewood: Business One Irwin, 1991): 8.

38. *The Texas Mathematics Teacher Bulletin*, Austin: University of Texas, 1920: 14.

39. Wrege, 97; also J.A. Hobson, "Scientific Management," *Sociological Review*, 4, 3, (1913): 197-212.

40. Simon Head, *The New Ruthless Economy: Work and Power in the Digital Age*, (Oxford: Oxford University Press, 2003): 6.

41. Walter B. Tardy, "Scientific Management and Efficiency in the United States Navy," *Engineering Magazine*, 15, 4 (July 1911): 553; also, *Literacy Digest*, 44, 16, (April 20, 1912): 840.

42. John C. Wood, Michael C. Wood, F.W. Taylor, ed., *Critical Evaluations in Business and Management*, (New York: Routledge, 2002), Peter B. Petersen, "The Pioneering efforts of Major General William Crozier (1855-1942) in the Field of Management," p. 120-121. See also *American Federationist*, April 1915, 292.

43. William M. Tsutsui, *Manufacturing Ideology: Scientific Management in Twentieth-Century Japan*, (Princeton: Princeton University Press, 1998): 32; also Sang M. Lee, Gary Schwendiman, *Management by Japanese Systems*, (New York: Praeger, 1982): 44-45; also *American Federationist*, April 1915, 292.

44. Peter J. Albert, Graig Palladine, ed., *The Samuel Gompers papers; The American Federation of Labor at the Height of Progressivism, 1913-17*, (Urbana: University of Illinois Press, 1986), Volume 9, 229; also U.S. Congress, House of Representatives, Hearings before the Committee of Labor, House Resolution 90, 62nd cong. 1st. sess. 1911; Testimony of Samuel Gompers, 22-25; U.S. Congress, House of Representatives, "Method of Directing the Work of Government Employees," Report No. 698, 66th Cong., 1st sess. May 17, 1916, 2.

45. Report No. 698, 2.

46. Leonard D. White, *The Republican Bra: 1869-1901; a Study in Administrative History*, New York: Macmillan, 1958, 173. In the 1880's, Navy Secretary William E. Chandler, received the following note from Philadelphia's City Committee
"Do not take away a large part of our party machinery by closing the Philadelphia Navy Yard. Nearly 400 of our active and useful committee men are employed in the yard and no one knows as well as you the power of patronage."

47. Frank B. Freidel, *Franklin D. Roosevelt: The Apprenticeship*, (Boston: Little Brown, 1952): 193.

48. Morrison, 56.

49. Congressional Record, 64th Cong., 1st sess., Vol. 53 (March 21, 1916) 436 (Senator Harding notes that navy deficiency budgets cover yard losses). Also J.A. Stockfisch, *Plowshares Managing the American Defense Establishment into Swords*: (New York: Mason and Lipscomb, 1923), 55.

50. Morrison. Daniels replied, "I will appoint you a commander in the reserve force and give you the job as one of Uncle Sam's naval officers." 87.

51. U.S. Senate Congressional Record, March 21, 1916, p. 4536, remarks of Senator Warren Harding.

52. John Douglas Forbes, *Stettinius Sr.: Portrait of a Morgan Partner*, (Charlottesville: University Press of Virginia, 1974), 65; also Ron Chernon, *The House of Morgan: An American Banking Dynasty and the Rise of Modern Finance*, (New York: American Monthly Press, 1990), 188-191.

53. Chernon, 189.

54. John M. Blum, *Joe Tumulty and the Wilson Era*, (Boston: Houghton Mifflin, 1951), 142.

55. Roderick L. Vanter, *Industrial Mobilization: The Relevant History*, (Washington, D.C.: National Defense University Press, 1983), 5.

56. Fleming, 107.

57. Morison, 145; See Fleming, 307.

58. Tracy Barrett Kittredge, *Naval Lessons of the Great War*, (Garden City: Double Day Page, 1921): 325-326.

59. U.S. Congress, House Committee of Naval Affairs, "Naval Appropriation Bill, "66th Cong., 1st Sess. June 19, 1919, House Res. 5601, Part 2, 100 Daniels testimony "We must have one control of wireless in America. It must be the government..." also David Kennedy, *Over Here: The First World War and American Society,* (New York, N.Y.: Oxford University Press, 1980):362.

60. Susan J. Douglas, *Inventing American Broadcasting, 1899-1922,* (Baltimore: Johns Hopkins University Press, 1987), 281. Philip T. Rosen, *The Modern Stentors: Radio Broadcast and the Federal Government 1920-1934,* (Westport; Greenwood, 1980), 10.

61. L. S. Howeth, *History of Communications-Electronics in the United States Navy* (Washington: Bureau of Ships, 1963):356.

62. Kenneth Bilby, *The General: David Sarnoff and the Rise of the Communications Industry*, (New York: Harper and Row, 1988): 52.

63. Faulkner, 86: Peter Fearson, *War, Prosperity and Depression: The U.S. Economy, 1917-1945*, (Lawrence: University Press of Kansas, 1987), 86.

64. Freidel, 35

65. Frank Freidel, *Franklin D. Roosevelt: The Apprenticeship*, (Boston: Little Brown, 1952), 201.

66. Naval Administration, Selected documents on navy department organization, 1915-1940, 19.

67. O'Connell, 60.

68. Daniel M. Smith, "Robert Lansing: 1915-1929," Norman A. Graebner, ed., *An Uncertain Tradition: American Secretary of State in the Twentieth Century*, (New York: McGraw-Hill, 1961) 123.

69. Edward S. Miller, 113

70. *New York Times*, June 5, 1920, 1; also December 17, 1920, 16.

71. "If America and Japan Went to War – A Japanese's View," *Literary Digest*, 67, 11 (December 11, 1920): 52.

72. Hector L. Bywater, *Sea-Power in the Pacific: A Study of the American-Japanese Naval Problem*, (Boston: Houghton Mifflin, 1921), 16. Also NARA, RG 80, Secret Correspondence, General Board to Secretary the Navy War Portfolio, Orange War Plan, March 14, 1914, Box 6

73. William Reynolds Braisted, *The United States Navy in the Pacific, 1909-1922*, (Austin: University of Texas Press, 1971) 510-511.

74. NARA, RG 127, Records of the U.S. Marine Corps, *War Plans and Polices*, War Plans, 1915-1946, Box 3,General correspondence pertaining to war portfolio, memo from Chief of Naval Operations to Major General Commandant, U.S. Marines, January 28, 1920, Marine Corps Archives, Breckinridge Library, Quantico, VA.

75. NARA, RG 80, Roll 64, Fourteenth Naval District, U.S. Naval Station, Pearl Harbor, Commandant to Director of Naval Intelligence. Subject, H.I.J.M.S Yakymo, March 16, 1920.

76. Braisted, 546; Carl Boyd, Akihiko Yoshida, The Japanese Submarine Force and World War II, (Annapolis: Naval institute Press, 1995) 14; Also *Army and Navy Journal*, 576, April 10, 1920, 966.

77. NARA, RG 80, General Board, Pearl Harbor, March 3, 1920, 4.

78. NARA, RG 80, Confidential Correspondence, memo from Chief of Naval Operation to War Plan distribution list, Subject: Marine detachment at Pearl Harbor, August 11, 1920.

79. NARA, RG 80 from Commander-in-Chief (Admiral Albert Gleaves) to Secretary of Navy, 8 December, 1920.

80. NARA, RG 80, Commander-in-Chief, Secret Order, *Special Readiness Report*, U.S.S. Swallow, Bremerton, Washington, January, 1921. Also Edwin Denby papers, Robinson Deposition, 69. Detroit Public Library, Burton Historical Collection, Detroit Michigan.

81. NARA, RG 80, Confidential Correspondence, Memo from the Board for the Construction of Navy yard Plans to the Secretary of the Navy, subject, *Pearl Harbor* November 3, 1920.

Chapter Two

The Harding Administration, 1921-1923

ADMINISTRATIVE OVERSIGHT

Warren Harding assumed the presidency on March 4, 1921 and served until August 2, 1923. The President had returned from Alaska aboard the navy transport Henderson when he died suddenly in San Francisco. We begin by examining the general orders, executive orders and naval policies of the Harding Administration.

President Bureau of the Budget

Harding had campaigned on the need to control federal spending. After Congress passed the Federal Budget and Accounting act in 1921, Harding offered Charles Dawes, a Chicago banker, the post of Treasury Secretary. As an army Brigadier General, Dawes had managed General John J. Pershing's supply activities during the First World War. To Pershing's eternal gratitude, Dawes unraveled the army's inventory snarl in Europe. The experience left a mark on Charles Dawes, however. If the military, an organization of discipline and command, had experienced inventory problems, how did the government's nonmilitary side acquit itself?

Dawes was unusually blunt in testifying before Congress. Lawmakers asked if the Army had overpaid for its war material before a congressional committee in February, 1921. Dawes replied; "I would have paid horse prices for sheep, if sheep could have pulled artillery to the front."[1]

In any event, Dawes turned down Harding's offer on grounds that the Treasury Secretary had insufficient authority to cut government spending. Harding asked Dawes to run the new Bureau of the Budget. Dawes agreed to serve as director subject to two conditions; he would serve no more than one year; he would report to the President and no one else. Harding agreed.

Dawes next assembled 600 federal officials in the Interior Department's auditorium. He reminded them that, rather than representing their own agencies, they were obligated to serve the office of the President. Dawes appointed a budget officer in each department who reported directly to the Budget Bureau. An officer's performance would be measured by their ability to deliver cost reductions.

Dawes then held up an Army broom in one hand, a Navy broom in the other. The Navy, he said, had recently ordered 6000 new brooms and paid $5.00 for each. The Army had 5,000 brooms in storage. The conclusion was inescapable. The Navy could have acquired 6,000 brooms "free." Dawes insisted that the Budget Bureau permitted the President to view government operations as a whole and thus set spending priorities. A former banker, Dawes sought to transform the president into a chief executive officer. [2]

Over the next several months, Dawes briefed the President on the progress of his spending reduction. Dawes recorded that each session took less time. The President, he said, anticipated, understood and approved the director's proposals. Warren Harding was apparently conversant with the concept of revenues, expenses, deficits and debt. And Dawes did deliver. Within a year Dawes had reduce federal spending by over billion dollars. [3]

Any government reduction program was not totally leak proof, however. An executive agency could slip a spending request to a member of Congress who appended the bill to larger appropriations—a back door process that circumvented the budget office's vetting system. Overall, however, Dawes cost reduction prepared the way for the administration's next step, tax reduction.

Secretary of the Navy

Harding asked John Weeks, a former Massachusetts Senator, to serve as Secretary of the Navy. An Annapolis graduate, Weeks knew many of the department's officers first hand. Weeks informed the president that, as secretary, he would be placed in a conflict of interest. Harding offered and Weeks accepted the cabinet post of Secretary of War. The President asked if Weeks had any suggestions for the Navy portfolio. Weeks mentioned a former house colleague, Edwin Denby. Denby was offered the Navy secretariat and agreed to serve.

Edwin Denby had enlisted in the Navy on two occasions. During the Spanish-American war he served as a gunner's mate. In the First World War, Denby enlisted in the Marine Corps. Between the wars Denby entered politics and was elected three terms in congress, representing Detroit's first district. He was defeated in his fourth election attempt in 1910.

Denby began to assess his office's span of control. In a sense, the secretariat was over diversified. Denby decided to trim some activities. He transferred Naval coal property in Alaska to the Department of the Interior (Secretary Weeks, War Department, turned back its Arizona holdings to Interior.) Next, Denby found sealed oil bids on his desk left from Josephus Daniels' tenure in office.

Denby was informed that the Secretary's office operated the Navy department's oil desk. Before making any administrative change, Denby was aware of several policy realities. Congress had assigned him an office assistant and several messenger runners. Second, the Secretary's oil desk was staffed by two officers on loan from the Bureau of Engineering. Third, Denby was informed that private drilling was tapping into adjacent Navy crude oil. Fourth, the Navy Secretary was responsible for 45,000 acres of petroleum reserves; Interior's Bureau of Mines and the United States Geological Survey managed 17.6 million acres of public land. Interior employed over 100 specialist and experts' familiar with reserve oil estimates and royalty calculations. Naval officers assigned to navy oil reserves were essentially trained in fleet command. Finally, President Harding asked his cabinet members to identify and eliminate overlapping government operations. In Denby's mind two executive departments running the same program represented a classic case of administrative overlap. [4]

The Secretary asked Albert Fall, Interior Secretary, if he would administer reserves 1, 2, and 3 on behalf of the Navy Department. Fall appeared hesitant to do so, but acquiesced to Denby's request. Denby next met with President Harding and informed him of his proposal to employ Interior as a Navy agency, noting the duplicating functions of each. The President, working on an executive reorganization plan, agreed that one department alone should be responsible for land administration. In May, 1921, Warren Harding signed an executive order. Under the order, the Navy retained oil reserve ownership.

Rear Admiral Robert Griffin, Chief of the Bureau of Engineering, took exception to the President's order. Griffin wrote a memo outlying his opposition. Denby appended the Admiral's reservations in a letter submitted to the President. Assistant navy secretary Roosevelt Jr., hand delivered Denby's order to the White House. Griffin's appendix was somehow mislaid. It was later found and made available to Congress and the public. [5]

In October of his first year in office Denby transferred the Secretary's oil desk to Rear Admiral John K. Robison, Chief of the Bureau of Engineering. Denby directed the Bureau Chief to act as the navy's liaison with Interior. Denby also instructed Robison to ensure that Interior's oil reserve assignment supported the Navy's war plan, a plan that contemplated two ocean war scenario. [6]

After Denby delegated coal and oil reserve management to Interior, the Secretary turned to the Navy's shore establishment. The shore establishment dealt with material supply, ships and ordnance. In this arena the Secretary uncovered several administrative problems.

First, Navy Yards were managed by line officers assigned temporary shore duty while awaiting duty afloat. Trained in fleet command and tactics, most Commandants were neither schooled nor experienced in naval construction. Moreover, line officer rotation resulted in management discontinuity.

Second, a yard Commandant managed a range of civilian trade specialties. Yard employees were naturally interested in job security. Over time, Congress permitted Navy Yards to diversify into vessel components as well as ship construction. Indeed, Congress assigned construction work to Navy Yards if they experienced excess capacity.

Third, Bureau personnel assigned to yard duty ostensibly reported to the yard Commandant. In practice, Naval officers identified with their respective bureau chiefs, each of whom was responsible for a particular ship component. Any Commandant found it difficult to orchestrate inter bureau coordination inherent in any warship construction. Multiple bureaus generated a blizzard of invoices for ships under construction.[7]

Fourth, the Navy's Bureau chiefs ostensibly reported to the CNO and the Navy Secretary. In practice, neither the CNO nor the Secretary controlled bureau spending. That authority rested with Congress. Predictably, the Bureau Chief's expended time and energy cultivating an intimate relationship with members of the House and Senate Naval Affairs Committee. In the give and take between Bureau Chief's and Congress, it was not surprising that political patronage prevailed over competence and merit.

In September, 1921 Denby issued General Order #68 setting up a Navy Yard Division in the secretary's office. The General Order unified the navy's yards under a civilian secretary. Next, Denby directed the office of the assistant secretary to serve as general yard oversight. The General Order instructed that office to do the following;

> cost analysis in connection with determination of economy and efficiency; supervise civilian personnel and labor, administer plants, machine tools, equipment and appliances, coordinate funds for yard vessel work; inspections of navy yards and stations.[8]

Denby's General Order was striking in content and thrust. In content the Secretary intended to monitor the economic performance of the navy's shore establishment; in thrust the secretary focused on cost accounting and efficiency. Congress had banned and prohibited government plants from adopting the basic principles of scientific management. The secretary had chosen to walk into the lion's den.

Denby next added a new administrative office to the navy's yard operation—an industrial manager. The manager, experienced and knowledgeable in ship construction, would take over the responsibility of vessel building and repair work. The industrial manager would confer an element of administrative continuity to yard operations. In short, the secretary attempted to unify and standardize the navy's shore establishment. Perhaps then a Navy auditor could have an index that measured differences in yard performance.

Assistant Navy Secretary

In addition to yard oversight, Denby assigned another matter to the Assistant Secretary: naval disarmament. In the summer of 1921, President Harding and Charles Evans Hughes, Secretary of State, announced a naval disarmament conference scheduled for November 1921, Washington, D.C. In preparation for negotiations, Secretary Hughes asked the Navy's General Board to submit the battleship tonnage essential for defense purposes. The board came up with one million tons. Hughes requested a second estimate. The board reduced the number to 800 thousand tons. Hughes still remained dissatisfied. Denby bypassed the board and appointed three advisors to the Secretary of State, Admiral Robert Coontz, Admiral William Pratt and Theodore Roosevelt, Jr. [9]

Denby and Weeks next addressed the issue of mobilization. Both secretaries agreed that the frenzied resource bidding of the First World War had resulted in production bottlenecks that spilled into price inflation. With that experience in mind, the two Secretaries proposed an Army-Navy Munitions Board (ANMB). By creating a coordinating body, both Secretaries attempted to anticipate military material requirements before rather than after the outbreak of hostilities. Theodore Roosevelt, Jr., the assistant secretary, now added industrial mobilization to his overall portfolio. [10]

In mid-1922, Denby had to staff the Naval Research Laboratory. Although Congress had appropriated a million dollars for buildings and facilities, the laboratory was without an operating budget. Josephus Daniels had also left the personnel composition of the laboratory up in the air. The issue rested with Denby.

The Secretary pulled together several activities. He acquired the Naval Research activities from the National Bureau of Standards, Department of Commerce. He brought in the Aircraft Radio Laboratory located at the Anacostia Naval Station together with a Sound Research Laboratory from Annapolis. Finally, Denby incorporated a Heat and Light division. The Secretary, in short, assembled scattered research activities and placed them under one administrative roof. [11]

Denby was aware that Thomas Edison had advised against placing a naval officer in charge of the laboratory. Edison felt that a military head would inhibit the laboratory's creative impulses. Edison recommended that the lab's research effort should be placed under civilian management. Several officers, on the other hand, felt the new lab should be a wholly owned subsidiary of a Bureau. Between these extremes, Denby struck a compromise. He placed the laboratory under the Office of the Secretary, assigning the Assistant Secretary, a civilian, as administrator. He then appointed a Navy captain to serve as laboratory director. [12]

Denby adopted an interdisciplinary model somewhat analogues to a university research setting. In describing the laboratory's culture, one historian observed: "...the atmosphere at the station was clubby, academic and conducive to productive study". The laboratory, nevertheless, elicited few financial patrons. Rear admiral John K. Robison, Engineering Bureau Chief, intrigued by the possibility of submarine detection, provided research funds in the lab's early years. [13]

Office of the Chief of Naval Operations

Denby's administrative changes reached into fleet readiness. Although, the CNO was responsible for war planning, Congress had given the office no direct authority over the department's bureau system. The Secretary decided it was time to elevate the authority of the Navy's ranking admiral. Accordingly, Denby designated Admiral Robert Coontz, CNO, as the department's budget officer. The move stirred a response. The Bureau of Supplies and Accounts, the navy's pay master, objected to the appointment, asserting that the CNO impinged upon the Bureau's independence. More important, the Navy's budget officer resided between the Bureau Chiefs on one side and members of Congress on the other. It would not be long before House members would call the secretary's administrative move to account. [14]

Denby's next turned to Operation's War Plans Section. Here the Secretary encountered another case of administrative overlap. Congress had assigned two agencies to war planning: the General Board and the Operations office. Predictably the Board and Operations did not always agree on the navy's strategy in the Far East. Operations, for example, concluded that the Philippines islands, in any war with Japan, were expendable. The General Board, by contrast, insisted that the defense of the Philippines islands was critical to U.S. Pacific interests. Theodore Roosevelt, Jr. recorded in his diary that Operations and the General Board continued to step on each other's toes. The resolution rested with the Secretary. Denby proceeded divest war planning from the General Board and elevated the stature of War Plans in Operations. [15]

Denby inherited Daniels' war plan orange (Japan) in the Pacific. Within less than a month in office, the secretary signed off on the plan. That plan defined Pearl Harbor as an advance base in the Pacific. In this arena the secretary had a run of good fortune. Rear Admiral Clarence Williams took over as head of operation's war plans in the summer of 1921. Williams possessed a gift for strategic insight and naval planning. In the next year and a half, he would issue memorandum, suggestions, comments covering naval oil reserves, oil depot sites, fleet organizational structure, command protocol, amphibious operations, marine lighters, the Navy's Transportation Service, logistics afloat, a merchant marine subsidy program. Little escaped Williams' wide angle lens. Later as president of the U.S. Naval War College, Williams continued to state his views on matters of fleet readiness. [16]

The Secretary's next turned to the matter of cost savings within the Navy Department. He appointed a committee to make recommendations. The committee quickly bogged down on the CNO's authority over the department's Bureau Chiefs. Operation's personnel sought to expand the CNO's cognizance. The Bureau Chiefs countered that line officers were ill qualified to make technical and engineering judgments. One officer proposed a quasi-staff system as a compromise. The proposal fell apart over funding. Congress had prohibited any Secretary from transferring monies between or among the Bureaus. Even if they so desired, the Chiefs insisted that Congress had banned inter bureau financial transfers. The committee reached a stalemate. Denby informed the committee that he had hoped they would combine the Bureau of Engineering and Construction and Repair into a single bureau as a coordinating move. In his first year in office, the Secretary did not press the matter. [17]

The Secretary did receive support from an unexpected source, the Bureau of the Budget (BOB). In evaluating the Navy Department's administration, budget officials noted that the Bureau chiefs not only ran their own multimillion dollar operations, they also acted as advisors to the Chief of Naval Operations. BOB concluded that the chiefs' work load was over extended. Running their own bureaus was a job sufficient unto itself. The budget officer recommended that the Navy create a formal, administrative staff, not unlike that adopted by the U.S. Army. The budget officer put the issue squarely; no staff, no coordination. The BOB suggestion found few supporters. It was clear that Congress preferred to act as bureau coordinator. [18]

Four years after the war, President Harding continued to demobilize the nation's war effort. From a budget of two billion dollars in 1920, the navy's annual expenditure approached $300 million by 1923. The fleet's fuel budget was, in fact, so tight that the CNO had to cancel annual fleet maneuvers. And few administrative details were overlooked. The Navy rationed long distance telephone calls and ordered its staff to reuse carbon typewriter ribbons. [19]

In July 1923, Denby kicked an administrative hornet's nest. Admiral Robert Coontz's tenure as CNO was about to end and he anticipated becoming Commander-in-Chief of the U.S. fleet (CINCUS). Admiral Edward Eberle, formerly CINCUS, was scheduled to succeed Coontz as CNO. In

mid-July the secretary informed his council that he proposed to give the
CNO coordinating authority over ship repair and ship alteration as part of the
fleet's readiness mandate. Denby's order attempted to reign in three bureaus
responsible for the bulk of the department's spending.

Denby announced his proposed change at July's Secretary's council
meeting. The Bureau Chiefs in attendance included; Rear Admiral C.B.
McVay (Bureau of Ordnance); Rear Admiral John K. Robison (Bureau of
Engineering); Rear Admiral Julian Latimer (Judge Advocate Office); Major
General Feland (Marine Corps); Rear Admiral David Potter (Bureau of Sup-
plies and Accounts.) The following discussion reveals the council's reaction;

> Admiral McVay: The Chiefs of the Bureau of Engineering and Ordnance
> are already carrying that with the Chief of Naval Operations. This would
> only put in the regulations what we are already doing.

> The Secretary: Don't you think it is desirable? And if so don't you think it
> should say so in the regulation?"

> Admiral McVay: Yes. To supply, yes, if it means what it says, it is an
> excellent thing. If it means the functions of the bureau are to be taken
> over by someone else; it is objectionable.

> The Secretary: Obviously the commissioning of a ship is the duty of
> Operations. Operations communicate with Engineering, C&R, Ordnance
> and Supplies and Accounts. S. & A. have your supplies aboard at a certain
> time, in other words, coordinating the preparations in the provisioning of
> that ship and so with other things pertaining to the operating of the Navy.
> The conducting of the fleet always should be vested in Operations, who
> should have authority to direct how and when the material, supplies,
> personnel of a ship shall be laid down. It is not in the regulations. It
> should be in the regulations. It doesn't deprive a Bureau Chief of author-
> ity at all. There is nothing now in it.

> Admiral Robison: The language is a little broader than the present prac-
> tice, and while it is undoubtedly wise and necessary, the language might
> well cover practices broader than—I wouldn't be asked to receive orders,
> for instance, from a subordinate of the chief of operations. I don't think it
> would be proper and legal.

> The Secretary: It depends upon the mere mechanics of the thing. The
> Bureau Chief is responsible in his department. The Chief of Naval Opera-
> tions is responsible in his department. If the order is proper it should be
> obeyed whether it comes through a subordinate or the Secretary of the
> Navy. The moment the bureau chiefs find that the Chief of Naval Opera-

tions issues an improper order, and then he takes it to the Secretary. It doesn't in any way restrain or restrict the Bureau Chief from coming to the Secretary with a protested point.

Admiral Latimer: My understanding is that it does not restrain in any way the functions of a Chief of Bureau.

The Secretary: Nor force a Bureau Chief to do anything he considers illegal. It doesn't in any way abridge the authority of the Bureau Chief.

General Feland: Admitting that we are doing that now, what is the reason for changing the regulations?

The Secretary: We have a very splendid co-operation today this is admittedly the thing to do. It should be continued as matter of regulations because next year or a few years hence you will have a new Secretary and a new deal all around.

General Feland: We are doing that very thing now. Admiral Robison brings up the question of the legality. The present regulation as I understand it is based on law. We are writing into the regulations something which may or may not be legal.

The Secretary: If we are doing that which is illegal now, should not continue, but if it is lawful it should be in the regulations.

General Feland: I don't mean unlawful, but I mean based strictly on the reading of the law.

The Secretary: I am trying to make no changes in the handling of personnel, for instance. What I am trying to do is to write something in the regulations to guide the future secretary, unless he wants to change the regulations. It is always a difficult thing to change the regulations.

Admiral Latimer: I wouldn't submit a regulation that I didn't think was legal.

Admiral Robison: It makes a rather radical change in the legal organization of the department. It doesn't change in the least, as I understand it, the present administration of the department. The present administration of the department is accomplished through the love of the service and you are changing that, to accomplishment by virtue of the law of force. It may be wise. I don't know. It changes in the least the way I do my business. I am a little doubtful. I will do more for love than from force.

The Secretary: I think it is wise, if it is something we know is good and successful, to make it a custom. This arises simply because of the splendid coordination which has been going on so satisfactorily that I don't want to see it discontinued. That is not any possible a reflection upon Admiral Robison. The service will remain just the same. The practice is now in force, the change is a trifling one.

Admiral Robison: No, it is a great change. It may well be a desirable change.

The Secretary: All of you know without my pointing it out that the system you operate under now has always been in vogue, and you have pointed the way to cooperation.

Admiral Robison: We have not done it ourselves. We have—I mean we have all tried to cooperate.

Admiral Latimer: It is now in the regulations who shall advise Secretary in regard to military features, etc., and also as to all matters pertaining to fuel, reservations, targets, radio stations, fuel, stores and other supplies with a view of meeting effectively the demands of the fleet. In preparing and maintaining plans he shall consult with and have the assistance of various bureaus and so on.

Admiral McVay: That is what we do now.

Admiral Robison: I know I have detailed myself as a member of the planning committee; every bureau has a representative there. I think that is a necessity. Potter has had an officer for a considerable time detailed up there just for that purpose so that our plans may be reasonably possible of execution. That is the only way in the world you can put into the plans those concrete things upon which success has got to be based.

The Secretary: Any further comment on this? …Then this will go to the President and we will pass to another matter. [20]

Edwin Denby incorporated the CNO's authority under General Order 433. Whether he anticipated any reaction from congress is not clear. But General Order 433 was certain to impinge upon congressional sensitivities. In a matter of time, Denby would hear from Capitol Hill.

Division of Fleet Training

On June 6, 1923, Denby added another division to the Office of the Chief of Naval Operations, a Division of Fleet Training. His General Order specified the following:

1. The Division of Fleet Training will be coordinate with the other divisions of the Office of Naval Operations, and will be under the charge of a director of the rank of rear admiral or captain.
2. The present Division of Gunnery Exercises and Engineering Performances are discontinued and its function is incorporated in the Division of Fleet Training.
3. The Division of Fleet Training, under the Chief of Naval Operations, is charged with

   a. The preparation of a balanced program of fleet training based upon approved war plans, and the current degree of readiness for war in each branch of fleet training.
   b. Cooperation with the fleet and the Naval War College in study, research, and experiment in all branches of fleet training for war.
   c. The preparation of general instructions for the conduct of fleet exercises.
   d. The preparation of the War Instructions, and other standard instructions, manuals and drill books governing the military activities of the fleet in war and in training for war.
   e. The collection, analysis, and review of all data in regard to fleet training and compilation of the same into suitable reports for the information and guidance of the service.

   In order that the Division of Fleet Training may efficiently perform its duties, it will maintain close contact with the fleet all times and will be conducted with a view to providing plans, standard instructions, and permanent records in all branches of fleet training.[21]

During annual exercises, the fleet participated in gunnery and engineering competition, a financial award given to ships and personnel for outstanding performance. Still, Denby's Fleet Training Division differed from past exercises. First, his general order emphasized fleet tactics . Second, the Atlantic and Pacific fleet, now designated the U.S. fleet, participated in maneuvers as a concentrated force. Third, although CINCUS remained in command of the fleet, the Commander-in-Chief had to clear his proposed exercise with the Chief of Naval Operations. The approval gave the CNO an element of oversight over fleet problems. Finally, Operations personnel served as umpires on such matters as ship damage, casualties, vessel loss, engineering awards. On more than one occasion line officers would take issue with umpire's calls. [22]

Working with War Plans, the Division of Fleet Training solicited officer reaction to publications that included radio manuals, gunnery rules, engineering instructions, tactical procedures. Subject to revision, the publications served as a guideline for future fleet maneuvers. In an indirect way, fleet publications took on the attributes of a feedback mechanism that enabled the CNO to exercise some influence over officer training and education.

Denby supplemented department coordination via a series of committees. He placed a War Plans Section into the Bureau of Construction and Repair, an addition assigned the task of converting merchant ships into fleet auxiliaries. Denby detailed a Marine to Operation's War Plans Section. Although John Lejeune was unable to place a Marine on the Army-Navy Joint Board, Denby did appoint a Marine on the Joint Board's planning committee, a subgroup that prepared working papers and agenda for the Joint Board. Denby designated a member of the War Plans Division and an officer from the Bureau of Yards and Docks to the Army-Navy Munitions Board, permitting the Navy to coordinate its supply requirements with that of the War Department. In the event of national emergency, the latter bureau would itemize the requirements essential for an advanced base. And Denby discretely assigned Rear Admiral Clarence Williams, war plans chief, to a presidential committee to promote a merchant subsidy program - a potential source of future fleet auxiliaries. In a final touch, Denby reinstated Secretary von Meyer's aid system by assigning an aid officer to the Secretary and the Assistant Navy Secretary. [23]

The Secretary thus employed formal and informal arrangements to rebalance the overall makeup units of the department's constituency. Each move appeared to be of little consequence. Viewed as a whole, they might someday exceed the sum of their individual parts. By 1923, the Office of the Chief of Naval Operations now approximated the following divisions:

1. Budget Officer
2. War Plan
3. Ship Movements
4. Naval War College
5. Intelligence
6. Inspection
7. Communication
8. Secretarial
9. Material
10. Naval districts
11. Fleet training

12. Naval reserves [added after 1923]
13. Reserve fleet [added after 1923]

Reserves: Personnel

Predictably, the Navy's reserve program expanded during the First World War. Following the Armistice, the nation's demobilization effort resulted in reserve officer and enlisted personnel reductions. Despite the cut back, a few reservists elected to remain with the service. Some regular officers were prone to view reservists as an impediment to their own promotion chances. Professor Ronald Spector put the issue in stronger terms, "...the regular officers viewed them (reservists) as a necessary evil and purged the service of most of them by 1921. Those that remained were not promoted."[24]

Naval aviation officers attempted to trim reservists from their ranks as well. Their technique was less confrontational—an examination. Appropriate for line officers, the test washed out reserve pilots. Still, intra-service tension paled in comparison to congressional budget cuts. It was Senate and House funding, or rather lack of spending that sounded the death knell for the navy's reserve program. The reductions were so severe that Denby decided to cancel the reserve program all together. The Secretary refused to let the concept wither away, however. Collaborating with the Judge Advocate's Office, the Secretary introduced a revamped program that included the Navy, Marine and, surprisingly, merchant marine personnel.[25]

At first glance, reservists appeared unrelated to the issue of fleet readiness. The Secretary insisted that once in a state of war, regular officers would be transferred to training rather than remaining on board an assigned warship. A reserve officer system, on the other hand, could relieve regular officers from instructional duty. Denby then transferred the Navy's reserve program from the Bureau of Navigation (personnel) to the Office of the Chief of Naval Operations. Dedicated to reserve training, he also placed the reserve fleet under Operations. The secretary attempted to move separate, bulkheads into a coordinated department.[26]

BALANCED FLEET

In December 1921, Denby reconstituted the U.S. operating fleet. First, he transferred the entire fleet to the Pacific coast—rescinding Daniels' early decision to split the fleet into Atlantic and Pacific squadrons. Second, he placed the fleet under one Commander in Chief, U.S. Fleet (CINCUS). Third, the secretary reconfigured the fleet's structure into a Battle Fleet, a Scouting Fleet, a Control Force, a Base Force and the Special Service Squadron. Battleships, cruisers, destroyers, and carriers were assigned to the Battle Fleet. The Scouting Fleet consisted of cruisers and destroyers; the Base Force included auxiliaries and the train; the Control Force made up of submarines, mining craft and amphibious support ships. Finally, the Special Service Squadron, a motley collection of older vessels, sometimes referred to as the state department's fleet, remained in the Atlantic to show the U.S. flag in the Caribbean and South America.[27]

Denby next adopted a fleet matrix or task force command structure. The task force model included warships assigned to a specific mission. A type ship commander, by contrast, assumed responsibility for ship maintenance and personnel training. Relieved from latter duty, a task force commander could reduce his staff and concentrate on fleet tactics and fleet training.[28]

The matrix format required a squadron commander to report to two superiors—a type commander and a task force commander. While participating in fleet exercises, a ship commander, reported to CINCUS. Once the exercise ended, individual ships reverted back to their squadron commander. The matrix/type fleet rotation became a rolling division of labor, enabling the task force commander to concentrate on tactical coordination.

Not every officer welcomed the matrix model. Admiral William Pratt, for one, favored the type command. In his view, a type commander encouraged officer overall responsibility. Denby's General Board #94 would remain in effect until 1930.[29]

Naval Arms Limitation Agreement

The Harding administration inherited Congress's Naval Act of 1916 legislation designed to give the fleet international supremacy. The officer corps obviously supported the fleet buildup. However, when the fighting ended in 1918, members of Congress began to harbor reservations over the cost of the 1916 program. A few members of Harding's own party began to take exception to Wilson's plan to build ten battleships and six battle cruisers. Indeed, before the President elect assumed office, Senator William E. Borah of Nevada called for an international forum to consider the question of naval disarmament. By early summer Warren Harding and Charles Evans Hughes, the Secretary of State formally invited France, Italy, Japan, and Britain to enter into disarmament discussions, scheduled to meet in November 1921 in Washington, D.C.

Britain hinted to the U.S. that its tradition fleet superiority was no longer beyond negotiation. Lord Lee, Admiralty head, informed Edwin Denby that Britain was willing to settle for naval parity with the U.S., a signal that Britain preferred to align its interest with the U.S. rather than Japan. The secretary passed the information to Charles Evans Hughes.[30]

During conference negotiations in 1922, British officers were startled by Hughes' insistence that Japan accept a tonnage quota inferior to the U.S. and Britain. Britain was unaware that U.S. naval intelligence had broken the

Japanese secret code and that secretary Hughes had knowledge of Japan's final bargaining position.

By February 1922, the Washington delegates announced the famous or infamous 5:5:3 warship formula, 500,000 tons allocated to the U.S. and British fleets; 300,000 tons for the Japanese fleet. The conference set a tonnage quota for aircraft carriers -146,000 tons for Britain and the U.S., 110,000 tons for Japan. Originally, Secretary Hughes wanted to ban the conversion of two battle cruisers into aircraft carriers. Theodore Roosevelt, Jr. made such a compelling case for the carrier that Secretary Hughes reversed himself. In the end, the conference permitted each party to convert two battle cruisers into carriers.[31]

It is easy to overlook what the Washington Limitation Agreement did not do. The agreement did not extend tonnage quotas to cruisers, destroyers or submarines. The conference place no restriction on the production of planes, auxiliaries, tenders, repair ships, transports, cargo vessels, stores ships, fleet tugs, landing craft or amphibious ships. Mobile dry docks, as a category, remained a questionable asset if self-propelled. On the other hand, the conferees did agree to a 10 year moratorium on battleship construction—a date later extended to January 1, 1937.

Japan accepted an inferior battleship tonnage on condition that Britain and the U.S. agreed not to fortify their western Pacific islands. To accommodate Japan, Britain transferred its Asiatic base from Hong Kong to Singapore; the U.S. stipulated that Pearl Harbor would remain its most western Pacific naval base. Some 3,500 miles resided between Hawaii and Japan. Most warships, at the time, had an operating radius of 2,500 miles. Japan signed the agreement on the assumption that its home islands resided beyond the reach of the U.S Fleet.

One agenda item remained an irritant to the U.S.: the Anglo-Japanese defense pact. The alliance held the risk that if Japan and the U.S. became belligerents, the U.S. might find itself exposed to a two-ocean war. The U.S. Navy sought to rescind the alliance—and the U.S. was not alone. Canada and Australia were uncomfortable with the alliance as well. In its place, the Washington delegates adopted a Five Power Treaty.

Japan agreed to remove its troops from China's Shantung Peninsula, and consented to drop the 1917 Lansing-Ishim's agreement that acknowledged Japan's special interest in China. The Washington treaty thus forestalled Japan's attempt to embark on its own Monroe doctrine in the Far East.

With few exceptions, U.S. Naval officer corps could not understand why the U.S. had abandoned Wilson's goal of fleet superiority. Josephus Daniels, was especially upset, noting that he had "sweated blood" to obtain congressional funding for the 1916 appropriation. Such criticism put Secretary Charles Evans Hughes on the defensive. The Secretary of State responded that he had checked to see if the U.S. Senate was willing to underwrite the cost of Wilson's battleships, much less upgrade Guam and the Philippines as Pacific bases. When Senator Henry Cabot Lodge responded that the Senate would refuse to do either, Hughes asserted that he had given away phantom warships and nonexistent bases in return for a naval construction moratorium. Rear Admiral John K. Robison, Chief of the Bureau of Engineering, put the matter in more stark terms. The Washington treaty, he said, had prevented a U.S. war with Japan, "...which was genuinely imminent of the time the conference was called."[32]

That observation, of course, did little to assuage naval professionals. Disheartened by the loss of warships, U.S. officers were confounded by the non-fortification clause that rendered obsolete a basic underpinning of their war plan orange strategy. In one sense, U.S. officers quarreled with not one but three presidents; McKinley for not buying the Micronesian Island; Wilson (Versailles Treaty) for permitting to retain Japan the islands; and Harding for defortifying U.S. bases in the western Pacific. Japan did agree to refrain from fortifying its newly acquired islands, but U.S. naval realists viewed that promise with cold skepticism.

Naval Aviation

Two months before the November Armistice Britain converted an Italian liner into the Argus - the world's first aircraft carrier. By the end of the World War I, Britain's Royal Naval Air Service (RNAS) stood in a class by itself, setting the standard in aviation, material, organization and operational experience. Trailing Britain, Daniels supported the conversion of a naval collier to the *USS Langley*, America's first carrier.

Before taking office, Harding had commissioned an advisory committee to examine the merits of aviation consolidation. The advisory group recommended that the Army and Navy retain and operate their own air service. The President included that proposal in an address before a special session of Congress on April 10, 1921. Thereafter, Theodore Roosevelt, Jr. and Edwin Denby drafted a General Order setting up a new Bureau of Aeronautics (BUAER). Admiral Coontz, supported the transfer of aviation from Operations to a Bureau, testifying that "I have not the time properly attended to the job," a somewhat unusual case when a naval officer declared that less was more.[33]

The new Bureau was not a spitting image of its sister Engineering, Construction and Repair, and Ordnance Bureaus. For one thing, aviationists preferred to buy their equipment from the private sector. For another, the bureau was committed to a specific power plant, the air cooled, radial engine. The engine possessed one overriding trait; it dispensed with water pumps, radiators and hoses—a displacement that increased engine reliability. The latter was critical. The ocean may have been an inviting shade of green but it was not a hospitable grass landing strip. General Mitchell, by contrast, favored the water-cooled power plant.

By consolidating naval aviation activities within one organization, Denby's corporate model was intended to quicken the bureau's decision-making cycle. Naval aviators were able to reduce the time required to get equipment approved. Flight officers could now define and develop their own tactical comparative advantage.

President Harding also rejected Mitchell's recommendation that the U.S. adopt a British Air Ministry that housed both commercial and military aviation activities. Rather, the President assigned commercial aviation development to Herbert Hoover, Secretary of Commerce. An aviation advocate, Hoover pushed navigational aid and safety rules as a vehicle to stimulate commercial aviation growth. General Mitchell, by contrast, discounted the prospects of U.S. commercial aviation, insisting that an aviation subsidy was comparable to two points of a railroad in a desert.[34]

Secretary Denby appointed Captain William Moffett as the Bureau's first aviation chief. A former battleship commander, Moffett had distinguished himself in the Spanish-American War. However meritorious the Congressional Medal of Honor, the award did not alter the fact that Moffett was not a trained pilot, a matter General Mitchell reminded Congress. The charge that the Bureau of Aeronautics Chief was a non-aviator must have stung. Admiral Moffett enrolled in the Navy's Pensacola Air Station and qualified as an aviator observer. (Later, the Congress inaugurated a policy that aviators alone could qualify as a carrier commander.)

General William Mitchell continued to appose naval aviation. He supported legislative that assigned all land bases to the army, a requirement that at first glance appeared innocent enough. Realizing the bill precluded the operation of naval air stations and flight training program, the Navy blocked Mitchell's congressional move.

In his early weeks in office, Denby attempted to convince Congress to fund the construction of a new carrier from the keel up. The war over, Congress was unreceptive to a large spending program. On this occasion, Theodore Roosevelt, Jr. made an audacious move. He convinced lawmakers to underwrite the expense of converting two battle cruisers into carriers by comparing the expense of a new carrier's cost against the cost of a converted battle cruiser. Roosevelt, Jr. argued that the latter would save the government twenty million dollars in construction cost per ship. Roosevelt, Jr., in a word, argued economy. Congress bought the proposition and in so doing gave birth to the carrier's Lexington and Saratoga.[35]

By the standards of its day, the carriers were one of a kind; a displacement of 33,000 tons (without blisters), a nine hundred yard flight deck, powered by electric turbines, a speed of up to 33 knots. Assigned to the Pacific's Battle Fleet in the late 20's, the carriers would participate in annual fleet problems. At the time the carriers were defined as a combatant auxiliary. Edwin Denby, nevertheless, speculated that the day might come when the carrier might embody an offensive capability all its own. Subsequent fleet exercises would put that notion to a test. [36]

In the early 1920's several students at the U.S. Naval War College became convinced that a circular formation had advantages over a rectangular disposition. The concept facilitated ship maneuvering while remaining in formation. Chester Nimitz convinced Admiral S.S. Robison to try the ring formation. Night maneuvers, however, proved another matter. By the late 20's, the ring formation had fallen into disuse.

AMPHIBIOUS WARFARE

Lejeune/Ellis

Edwin Denby had enlisted in the Marine Corps after the U.S. declared war in April 1917 despite his physician's concern that given his age and weight, Denby would be unable to handle the rigors of boot camp. Among other duties Denby was assigned as a Marine observer in France.

In a somewhat an awkward circumstance, Secretary Josephus Daniels had eased George Barnett out as Major General Commandant, Marine Corps, appointing John Lejeune as Barnett's successor. An acknowledged Democrat, a Republican Senate did not confirm Lejeune. Presumably, a new Navy Secretary, about to take office, would select his own Commandant. Before leaving the secretariat Josephus Daniels suggested that Denby retain John Lejeune as a Marine Corps head. At Harding's inauguration ceremony, Denby approached and asked Lejeune to remain as Marine Corps Commandant. Lejeune quickly won Senate confirmation. [37]

A Naval academy graduate, John Lejeune viewed the Marine Corps' future with some apprehension. Although the Marines had found a niche within the department, they also sought to distance themselves from the Navy, ever so slightly. To strike that balance Lejeune proposed to set up separate recruiting stations in government buildings. Such a move translated into cost savings. Although Denby bought the concept, he suggested that Lejeune clear the idea with Senator Miles Poindexter, Chairman of the Senate's Naval Affairs Committee. Poindexter signed on and Lejeune next took his plan to Budget Director Charles Dawes. Dawes exclaimed that Lejeune was the first government official to approach him with a cost-reduction proposal. To clinch the deal, Dawes handed Lejeune a cigar. Lejeune had moved one step toward Marine Corps differentiation. [38]

Lejeune was aware, nevertheless, that the Marine Corps remained vulnerable to "legislative death." The Marines' renowned struggle at Belleau Wood had, ironically, turned out to be a liability of sorts. Although the Marines'

Fourth Brigade had fought beside U.S. Army units, it was the Marines that garnered the glory and publicity. The war over, some members of Congress, bent on cutting programs, perceived the Marines as a service indistinguishable from that of the Army. As cynics put it, the Marines were really soldiers who spoke a peculiar Navy lingo.

Nor was it reassuring that the Marines had long served as gunner mates on U.S. warships. Some navy officers viewed the marine detachment as superfluous. President Theodore Roosevelt, in fact, attempted to ban Marines from battleship duty, a move blocked by Marine congressional lobbying.

Lejeune's colleague and friend, Lt. Colonel Earl Ellis, provided a solution to Lejeune's Pacific dilemma. Ellis and Lejeune's friendship dated back to the Philippines campaign following the war with Spain. Lejeune became Ellis's patron. When assigned to France, Lejeune took Ellis as his Adjutant. A Marine Corps Commandant Lejeune placed Ellis in charge of a Division of Operations and Training, an organization assigned to war planning.

Ellis had long been convinced that Japan and the U.S. were on a collision course. The fact that Japan had participated as an U.S. ally in the First World War did little to alleviate Ellis' concern. For one thing, the Japanese navy had trounced two Russian fleets in 1905; occupied Korea five years later; invaded China's Shantung peninsula; taken German held Micronesian islands in 1914, and attempted to control China's police force. Ellis concluded that Japan stood as an impediment to U.S. Pacific interests.

Prior to Lejeune's promotion as Commandant, Ellis labored on how the U.S. Navy could respond to a potential U.S.-Japanese conflict. Ellis anticipated that Japan could seize and occupy the Philippines and Guam, denying U.S. bases in the western Pacific. Ellis argued that the U.S. would have no choice but to take those bases by force. Ellis proposed amphibious assault as an offensive doctrine. [39]

Ellis did more. He assembled a time table and itemized the material requirements essential for any trans-Pacific operation. With Lejeune's support, Ellis fashioned a doctrine that complemented the needs of the Navy, yet set the Marines apart from the U.S. Infantry. Here the relationship between the Navy and Marines would be symbiotic—or at least that was the intention.

Not all Marine veterans were taken by Lejeune's vision of the corps. Having served in France, many were influenced by the Army's infantry weapons and ground tactics. A few went so far as to write off amphibious training as a waste of time. Lejeune and his staff tread carefully. Brigadier General Eli Cole, an amphibious convert, set up lectures during lunch breaks so as not to detract from the corps' conventional drill activity. Lejeune built upon Daniels' legacy of Marine Corps Schools by inviting British Gallipoli veterans to discuss their Ottoman battle experience, lectures supplemented by handouts, reading material, and books. The educational process would be an incremental one, and Marine acceptance of amphibious warfare would not occur overnight.

Blindsided by the Washington Naval Agreement, the Navy's officer corps laid out their own Japanese war scenario. If the two nations came to blows, the Battle Fleet would sortie from the U.S. west coast, refuel at Pearl Harbor, put to sea toward the Philippines, engage and defeat the Imperial Japanese Navy in a decisive confrontation. Under a Jutland scenario, the Marines would serve as occupation troops.

That narrative, of course, depended upon the U.S. Army holding the Philippines until the fleet arrived. As noted, the Navy's General Board and Operation's War Plans Division, did not see eye to eye on that issue. The latter was skeptical whether the fleet's "through ticket", to use Edward miller's term, was remotely feasible. [40]

What the Navy viewed as an obstruction, John Lejeune perceived as an opportunity. A few days after the Washington treaty was made public, Lejeune sent a letter to the Navy's General Board. The Commandant observed that the naval agreement did not preclude the formation of a mobile force. Lejeune submitted that a Marine Corps Expedition Force, (MCEF) would be that force. Such a force required transports, cargo, vessels, landing craft, beach parties, field artillery, and aviation units, to touch on the obvious. Lejeune wanted an expeditionary force to be self-sufficient. Lejeune proposed a matrix organization dedicated to carrying out an island assault. [41]

That Denby comprehended Lejeune's island hopping doctrine was evident in a confidential letter he sent to Senator Miles Poindexter, Chairman of the Senate Naval Affairs Committee in October 1921. Denby was obviously concerned about Hawaii's defensive position. He also appreciated Pearl Harbor's role in any Pacific offensive. Denby wrote Poindexter that any future Pacific war would reduce itself to a "contest of bases." The Marines, he said, proposed to take a hostile island, defend that island, and convert that island into a launching pad for subsequent island assault. In effect, Denby defined Pearl Harbor as an advance base 2200 miles west of the U.S. continent. [42]

The Marine Corps Expeditionary Force (MCEF)

Choreographing an ensemble of personnel and equipment marked the essence of any joint operation. The Marines anticipated that an enemy force would employ heavy artillery, aviation units and tanks in an island defense. Accordingly, the Marines sought tanks and tank training. Edwin Denby penned a memo to Secretary of War John Weeks requesting permission to send Marines to the army's Fort Meade tank school. The Army agreed to enroll enlisted and Marine officers. By 1923, the Marines had a light tank platoon. [43]

The Marines also wanted to deploy their own aviation unit. It was here that Marine pilots stood apart from their Navy brethren. For one thing, Marine's pilots were dedicated to ground support. To ensure that singleness of purpose, every pilot was required to serve two years with a ground unit. Only then could an individual apply for flight training. For another, a Marine could serve as pilot no longer than five years. After that, an aviator was reassigned to a non aviation unit. In short, the Marines defined an aviator as a Marine simply mounting a different weapon. No pilot was permitted to indulge in visions of bombing Tokyo.[44]

Harding's decision to forego a domestic RAF enabled the Marines to integrate aviation into their amphibious plans. To achieve close air support, Lejeune persuaded Congress to fund a Marine airfield next to Quantico, Virginia. Here resided a platform designed to give Marines an opportunity to work on aviation and ground coordination. By 1923, the Marines had an aviation training field. Later, Marine pilots, engaged in deep angle diving to achieve bombing accuracy. Denby, in the meantime, appointed a Marine Colonel, T.C. Turner, to serve as his aide. Turner was a Marine pilot.[45]

Control Force

As part of the fleet's reorganization, Denby established a Control Force assigned to support Marine and Army island assaults. The Control Force included a chief of staff, war planning, logistics, a flag secretary, a force radio officer, and a flag headquarters. Divided into task groups, the Control Force replicated the fleet's matrix command structure.[46]

The Marines realized that the success of any task force rested on unit coordination. Not surprisingly, the Marines were more than intrigued by wireless communications (radio). Field radios, of course, had become standard equipment in the Army. The Marines experimented with aircraft wireless linked to radio grounded sets installed in two and half ton trucks. Some officers speculated that radio might permit battleship artillery support in any island contest.[47]

Having outlined the weapons, materials and personnel associated with any amphibious exercise, Lejeune's next step was to find a suitable place for training purposes - a proxy for the Pacific tropics. A former lawyer turned Marine, Major Holland H. Smith, was assigned the task. Smith recommended Culebra and Viesques near Puerto Rico in the Caribbean. Here the Marines could engage in landing exercises in all its complexity. Holland Smith recalled later that the Marines placed a sketch of the Pacific over their Caribbean maps.[48]

Early exercises confirmed that landing craft, gun lighters, combat loaded ships, water-proofed radios were essential. The Marines preferred to employ the *USS Henderson* as a transport. Too often, the Henderson was booked, assigned to deliver troops or transport supplies. In the Henderson's absence the Marines relied on battleships as a substitute—though departing from a battleship and embarking on a destroyer proved awkward, laborious and time consuming.[49]

Landing exercises reaffirmed the need for specialized landing craft. Captain Ralph Earle, Control Force commander, wrote that standard navy launches broached, dumping Marines into the surf. After one such episode, Earle questioned why the Marines could not be delivered to an island shore "dry-shod."[50]

The Marines later experimented with an amphibious tank and a lighter capable of delivering a 155mm gun its associated tractor. The heft of the weapon, unfortunately, exceeded the boom capacity of most auxiliary ships. Despite the fact that a heavy warship could accommodate an artillery system the Marines continued to search for a more appropriate transport.

LOGISTICS

Ashore

War plans had long held that fleet logistics could not be divested from fleet readiness. Even before the First World War, the navy's General Board envisioned two defensive positions in the Pacific; an outer line that included Alaska, Pearl Harbor and the Panama Cana zone; an inner line that linked Bremerton, Pearl Harbor, and San Diego. In both cases, Pearl Harbor emerged as a strategic keystone.

In the early 1920's Albert Fall, Interior secretary, acting on behalf of the Navy, signed off on two oil naval leasing arrangements; one with Edward Doheny, head of the Pan American Petroleum and Transport Company; the other with Harry Sinclair, president of the Mammoth Petroleum Company. Doheny leased Californian reserve #1, paying a royalty between 30 to 50% of oil revenues to the U.S. Doheny agreed to issue royalty oil certificates in exchange for refined oil stored at Pearl Harbor and the west coast.[51]

Albert Fall insisted that he had originated the oil exchange concept. In a subsequent court deposition, Rear Admiral John Robison testified that he had conceived the royalty oil exchange. Actually, both were slightly off the mark. It was the U.S. Shipping Board, a legacy of the Wilson administration that first executed an exchange of royalty crude for refined oil.[52]

After the Armistice, the Wilson government encouraged the Shipping Board to compete in the Trans-Atlantic passenger market. Requiring cheap oil, the Board sought royalty crude oil from Salt Creek, Wyoming, a non-navy reserve just north of Naval Reserve #3, Wyoming. At first, Interior surmised that the two reserves were geologically separate and apart. It was discovered later that Salt Creek drained the Navy's #3 oil set aside. The finding was disconcerting. Two government reserves, one Navy, the other non-navy, were under oil lease arrangements. In effect, Congress permitted one executive agent to drain oil from another executive department.[53]

Interior negotiated the lease of reserve #3 with the Mammoth Oil Company. Among other requirements, the contract stipulated that Sinclair construct a pipeline just short of 1000 miles linking Wyoming to Missouri. The pipeline would then tie into U.S. refineries in western Indiana, linked to the east and Gulf coast. Interior estimated that reserve #3 held about 130 million barrels of crude oil. If the Teapot Dome turned out to be a dry hole, however, Sinclair's Mammoth Company could well take a financial hit.

Sinclair ran Interior's oil lease proposal by his board of directors. The board judging the venture too risky, rejected Sinclair's participation. Undeterred, Harry Sinclair formed a private company, the Mammoth Oil Company, and proceeded to sign Interior and navy's contract to place oil in 27 storage tanks scattered along the U.S. eastern shore.[54]

The navy's war plans called for the construction of oil storage tanks in the Panama Canal Zone as well. Denby wrote John Weeks, War Department, and asked if he would support funding fuel tank storage construction from the canal's transit receipts. Weeks responded that if the Navy supplied the oil and tanks, the War Department would provide the land.[55]

The navy's Pearl Harbor defense plan anticipated not only an oil tank farm but included channel dredging to permit battleship access to the yard's wharves and pumping stations. Although Pearl Harbor resided over 2,000 miles west of Bremerton, Washington, the Hawaiian base competed with bases on both coasts. Politically, it might have been tempting to route vessel overhaul work to the U.S. continent. Denby and Roosevelt, Jr. resisted that siren call. Instead, they ordered naval ships to direct their repair and overhauling requirements to Pearl Harbor as a means to support and sustain the base's civilian work force. Both secretaries had their eyes fixed on the Pacific.[56]

Advance Base

The Navy designated any base residing outside the continental U.S. as an advance base. To secure such a location, the Marines planned to seize an enemy island. Such an operation required auxiliary vessels converted from merchant hills. In a memo to Rear Admiral Clarence Williams, War Plans Division, Rear Admiral E.C. Kalbfus outlined the base requirements. The memo called for the following:

> ...erection of piers, wharves, roads, railroads, storage buildings and ships ashore to be provided with material for the rapid construction of piers, dredging, erection of marine railways, erection of store houses to release store ships, erection of ships for installation of such heavy armor-hull and machinery

repair tools as cannot be installed on repair ships, erection of barracks, fuel oil storage ashore.

Any Pacific logistics chain depended on a cooperative liaison between war plans, the Marine Corps and the Bureau of Yards and Docks.[57]

Mobile Base

Under a regime of tight budgets, the fleet attempted to save money by employing ship forces (crew members) for maintenance and repair work. As a supply back up, the fleet could also take advantage of repair ships or tenders. Maintenance work performed by ship forces or tender would enable the fleet to avoid continental yard overtime rates. Tenders held out a tactical bonus. The fleet could reduce their dependence on a fixed naval base. One option remained closed, however. Civilian work rules banned ship forces from engaging in repair work while their vessel remained docked at a Navy Yard.

Ironically, the Navy's cost reductions tilted the fleet toward tenders oilers, supply ships and repair vessels. As a move toward greater self-maintenance, naval warships were encouraged to expand their inventory of spare parts and supplies. Rear Admiral Robison, Engineering Chief, assured the Commander-in-Chief of the Fleet that he would supply spare parts to a maximum extent to secure a better service." Denby institutionalized fleet self-sufficiency by issuing General Order No. 2, directing commanders to "...affect routine maintenance and upkeep by ship forces [crew members] with the facilities on board plus those of repair ships..."[58]

On August 10, 1921, Theodore Roosevelt issued a memorandum to the Commander-in-Chief, Atlantic Fleet. Roosevelt ordered the Commander to determine whether the destroyer Shaw's maintenance work could be effected by "...ship forces or tenders..." Roosevelt Jr. also requested that a work report be sent to the Chief of Naval Operations.[59]

As a holdover from World War I, the U.S. Shipping Board owned a large inventory of oil tankers. In 1922, President Harding signed an executive order transferring seven tankers to the U.S. Fleet. Essential to any contemplated operation in the Pacific, a fleet tanker was more than auxiliary. A merchant tanker had the potential of evolving into a mobile Pacific base.[60]

As an economy measure, Secretary Denby ordered the Commander in Chief, U.S. Fleet to reduce the length of time a ship remained tied up at a Navy Yard. The order asked CINCUS to determine whether a warship upgrade could be accommodated by the existing ship forces or by resorting to a fleet tender. Again, both options moved the fleet toward a degree of fixed base independence.

Combined with fuel conservation measures, ship forces could prolong the time required before returned to a fixed base. Ship forces thus developed special repair and maintenance skills in a move toward self-sufficiency, supplemented by on board repair shops. With the addition of tenders and supply ships, the U.S. fleet began to enlarge its radius of action.

In the nineteen twenties the Navy reexamined the issue of fouled ship hulls. However prosaic, barnacle life cycles appeared worth exploring. Denby approved a grant to the Commerce Department's Bureau of Fisheries to study algae growth. The Navy also let a contract to the War Department's chemical warfare division in search of a toxic barnacle repellant. By the mid-twenties Norfolk Yard researchers had concluded that plastic paint embodied with cupric and mercuric oxide might extend the time a ship remained free from a dry dock visit. Antifouling paint began to take on the attribute of a force multiplier. [61]

Lawmakers asked Admiral Coontz why the Navy could not rely on commercial tankers for its refueling needs. Coontz replied that navy oilers required trained personnel and special pumps to permit simultaneous ship refueling at sea. At the time, replenishment underway employing port and starboard sides of an oiler may well have escaped public notice. [62]

Navy planners were vitally interested in merchant marine vessels whether operated by industrial fleets or as affiliates of the U.S. Shipping Board. The Fleet Training Division sought information on the per hour consumption of fuel oil. Operations also invited merchant fleet officers to attend the Navy's fuel oil center at the Philadelphia Navy Yard. If circumstances required that merchant ships be converted into fleet auxiliaries, at least they had exposure to oil conservation measures. [63]

In mid-1922, Admiral Coontz asked the General Board to lay out the requirements for a mobile base force. Coontz added that: "...such a base would be used as a substitute for a permanent advanced base in a campaign in the Western Pacific." Here the fleet made tentative moves in carrying its supply base with it, what some would later refer to as a fleet within a fleet. In his annual report, Admiral Edward W. Eberle, now CINCUS, noted that the use of tenders and repair ships fostered self-maintenance. Fueling at sea thus promised a triple return; it contributed to fleet mobility; it extended the fleet's operating range; it untethered the fleet from its fixed base dependency. Denby emphasized the essence of that dividend when he noted that "mobility" was the very soul of naval warfare. [64]

Logistics—Fleet Auxiliaries

The Harding administration also took steps to support the nation's merchant marine industry. Denby invited U.S. Shipping Board personnel as well as commercial engineers to engage in fuel conservation. Economy in commercial fleet operations was thus not totally unrelated to fleet logistics. [65]

Denby's refurbished reserve program included Navy and Marine officers. Surprisingly, the secretary's proposal covered merchant marine sailors and officers. The addition prompted U.S. trade union leaders to charge that the government could employ merchant reservists as strike breakers in the event of any industrial action. Herbert Hoover suggested to President Harding that retired naval officers should be encouraged to work for the U.S. Shipping Board if they chose to do so. To ease the transition from military to civilian status, Herbert Hoover suggested that the President exempt retired naval officers from civil service regulations.[66]

Harding addressed three merchant marine issues during his presidency. First, how should the U.S. demobilize the government's shipbuilding program dating back to World War I? Second, how should the government assist private ship operators? And third, how should the U.S. bolster the financial health of commercial ship building yards? As a first step, the President appointed Alfred D. Lasker, a Chicago advertising executive, to the U.S. Shipping Board. Lasker confronted a bulging inventory of merchant ships, some 200 constructed out of wood. He auctioned off the wooden vessels, then turned to the Board's current ship construction still operating around the clock. In an early morning visit, Lasker discovered that "...hundreds of men sleeping with nothing else to do." The chairman instituted a program of downsizing and layoffs.[67]

Lasker, with the President's back up, proposed a revamped merchant marine policy. He tied vessel speed to a U.S. mail subsidy. The subsidy was progressive; the faster the vessel, the higher the payout. Lasker also proposed financial assistance to the nation's private ship yard through tax breaks, low interest loans and insurance guarantees.[68]

Harding's ship program ran into congressional opposition almost from the start. Critics contended that the corporate fleets of U.S. Steel, Standard Oil, and United Fruit, by delivering products to themselves, qualified for a mail subsidy. Progressives in congress argued that big business hardly needed a federal bailout.[69]

The President neglected to point out that the federal government already subsidized government yards. Congress required that warship repair, modernization and overhaul work be restricted to government yards, demanded that warships be built at Navy Yards if they experienced excess capacity. Government yards were not burdened by taxes or construction bonds, had little accounting data on warship cost, were immune from cost over-runs, and incurred no penalty for tardy ship delivery. In fact, through the use of deficiency budgets, Congress set in place an incentive that promoted yard over-employment.[70]

After congress extended civil service status to yard civilian employees, due process became the venue dealing with personnel matters. Removing a shipyard employee for nonperformance or malingering became virtually impossible. Moreover, employee unions' pressured government yards to undertake work for the U.S. Shipping Board, the Army's Transportation Service, and the War Department's Light House Service. In diversifying Navy Yards into manufacturing vessel components, Congress responded to trade union pressure. Knee deep in political patronage, the House and Senate members, nevertheless, continued to insist that President Harding's merchant ship subsidy was essentially a raid on the nation's Treasury.[71]

Clearly, Harding's ship program embodied an element of national defense. For one thing fast merchant vessels could be converted into aircraft carriers, ocean liners and fleet transports. For another, merchant bottoms served as the fleet's supply train. When legislators questioned the navy's need of for auxiliary vessels, the Secretary responded that "...a ship may be in the highest degree a vessel of tactical value although not be able to fire a gun or torpedo." Denby also wrote that "...without the merchant marine to render mobile the necessary supplies of food, fuel, ammunition and repair, the navy is tied to its home bases," a veiled reference to Operation's mobile base plan.[72]

The ship subsidy bill managed to squeak by the House, although lawmakers disqualified corporate fleets from a post office stipend. The Senate proved even more difficult. Senator Robert LaFollete offered to vote for the bill if the President promised to reinstate a corporate excess profits tax. South Carolina Senators said they would add their support to the bill if the President reopened the Charleston navy yard. A Senate filibuster erupted and a cloture move fell short by one vote. The President's subsidy program expired. The upper body then took up a more urgent legislative matter—a bill that banned oleo margarine as a butter substitute. In 1923, consumer choice had not become a matter of overriding importance.[73]

That the President regarded his subsidy program as a matter of importance is suggested in letter he wrote to Senator Reed Smoot on November 4, 1922. The President said, "I am more deeply interested in the enactment of the merchant marine bill than any one measure I have ever recommended to congress." In defeat Warren Harding was eerily prophetic. Sometime in the future, he said, Congress would find it necessary to subsidize the nation's merchant marine industry.[74]

SOURCING, MAKE/BUY

Shore Establishment

The Navy addressed the sourcing options that had confronted the Wilson administration; namely, to secure equipment in house via its shore establishment, stations, factories or arsenals; or to purchase ships and hardware from firms residing in the nation's industrial sector. Prior to the First World War, private yards accounted for the bulk of navy combatant vessels. By the end of the war, a resuscitated shore establishment had altered that balance. The shore establishment had sufficient capacity to accommodate any "make" option with little difficulty.

The ground rules governing Navy and private yards differed slightly, however. Private yards had to submit firm prices; Navy Yards posted estimate costs. If the latter won the construction award there was no assurance that the final construction expenditure would approximate the original estimate.

Beyond any make/buy calculus, line officers remained less than confident as to the exact nature of yard costs. The give and take among the Bureau Chiefs within the confines of the Secretary's Council meetings suggested an element of frustration. Rear Admiral Robert Griffin (Engineering) on June 1921, observed that yard construction cost estimates were "fictitious;" Admiral C.B. McVay (Ordnance) concurred that Navy Yard cost estimates were "valueless." Admiral David Taylor (Construction) went so far as to assert that once a warship was completed, the Navy didn't know what it cost. The Bureau chiefs were surprised to learn that the Bureau of Supplies and Accounts, responsible for assembling cost data, had discontinued its work on grounds of a tight budget. More disconcerting, the Bureau of Supplies and Accounts had not bothered to tell the other Bureau chiefs of its decision. At one meeting, Argentina asked if U.S. navy yards could build a warship for its navy. The Bureau chiefs reviewed the merits of Argentina's case and turned down the request. U.S. commercial yards, concluded one officer, could do the job cheaper. [75]

That the Navy's accounting system accorded the yards considerable flexibility was evident when the Bureau Chiefs discussed work performed on a Brazilian warship. Before taking the job, the yards informed Brazil that overhead charges would account for 35% of the total expense. The job completed, the yard bumped its overhead cost up to 75% of the total—adding one million dollars to the final invoice. The Bureau chiefs discussed what appropriate action to be taken. They finally decided to submit the bill to Brazil and await the country's response. Perhaps the State Department could best handle the matter. [76]

In an environment of government downsizing, yard employees sent a delegation to the Navy Secretary's office to solicit government business from sister agencies within the federal establishment. Secretary Roosevelt Jr., however, questioned whether the navy's cost accounting permitted any accurate quotes at all. It was generally acknowledged that the navy's accounting system was tailored, not to the construction cost of a specific warship, but rather to satisfy congressional appropriation legislation.

The buy option, outsourcing in today's parlance, was not without its own problems either. Upon taking office, Edwin Denby inherited a contract dispute between the Navy and Electric Boat's engine affiliate, Nelselco. The Navy discovered that Nelselco's diesel engines could handle low speeds operations. At high speeds, however, engine vibration damaged boat drive shafts—a failure that, in the eyes of the Navy, pointed to Nelselco's flawed engineering design. [77]

The quarrel escalated to a point where the Navy withheld progress payments. Electric Boat countered by interrupted submarine delivery. Although Denby was able to broker a resolution between the two parties, the dispute apparently soured the navy on its outsourcing experience. At a Secretary's Council meeting Rear Admiral John Robison, announced the navy would, henceforth, confine submarine production within two government yards— Portsmouth, New Hampshire Navy Yard; and the Mare Island Yard, San Francisco, California. The New York Navy Yard would manufacture and supply required diesel propulsion machinery. [78]

During the First World War, the Navy operated a torpedo plant at the Newport Torpedo Station, Newport, Rhode Island, supplemented by a government plant in Alexandria, Virginia. The Navy also contracted torpedo production to the Brill Torpedo Company, a Connecticut firm. Under Denby's watch the navy cancelled the Brill contract, moth-balled the Alexandria operation and confined torpedo production to its Newport Station.

Outsourcing

Despite the navy's in-house aircraft factory, the Bureau of Aeronautics preferred to secure its planes and engines from private aircraft suppliers. Rear Admiral Moffett, in fact, provided seed money to the Pratt and Whitney Company as a device to stimulate engine competition with the Wright Aeronautics Corporation.

The Marine Corps naturally turned to the Bureau of Construction and Repair for their amphibious craft and equipment. But landing craft accounted for a relatively small portion of the navy's overall construction budget, and the material bureaus tended to assign a low priority to the Corps' somewhat off beat requirements. Major General John Lejeune, joined by Rear Admiral Ralph Earle, sought a self-propelled artillery lighter capable of handling a 150mm Howitzer and its associated tractor. The Bureau of Construction and Repair responded that a towed barge was the answer. The bureau's a response ignited a reply from Lejeune,

> It follows that the ideal characteristics of a lighter or boat suitable for transporting detachments of a landing force from the transport ship to a hostile beach are:
> a. Speed, in order to reduce the time under fire and to present a target difficult to hit;

b. Self-propulsion, in order (a) that it would not become a stationary target by reason of the disabling of a towing boat; (b) that it would not wallow around under fire after being "cast off" from a towing launch, and (c) that if not towed it could "zig-zag" and thereby present a target difficult to hit;

c. Unsinkability, if practical, in order to avoid drowning and to insure reaching the beach with its load. Interior division into wells would prevent a single shell sinking the lighter;

d. Overhead and frontal cover from shore missiles, in order to receive casualties;

e. Shallow draft, in order to permit a close approach to the beach;

f. Arrangements whereby the occupants could fire machine guns or 37mms, at the beach when within range, in order to remove the defenseless feeling rather than to inflict damage on the enemy;

g. Capacity – for 65 men, or 125 men; or 250 men. [79]

Lejeune suggested that a special procurement board be set up as an administrative unit to explore the equipment needs of the expeditionary force. The Bureau of Construction and Repair rejoined that it not only wanted to be represented on the board but designated a specific officer to fill that position. Over time, Congress had placed the bureau in the position of passing judgment on its competitor's products, an ambivalent status apparently regarded as a procurement nonevent. Still, Lejeune's request was suggestive. Given its unconventional material requirements, the Marine Corps began to look beyond the Navy's bureau system for its equipment needs. Frustrated by the navy's integrated structure, the Bureau of Aeronautics, submarine service and Marine Corps discovered they had much in common. [80]

Incentives

The Harding administration found it difficult to completely demobilize Wilson's war expansion. Muscle Shoals, for example, operated an electric power grid in Alabama as an investment in nitrate production. The war over, the administration put the facility and its dam infrastructure up for sale. Ford Motor Company offered to buy and convert the plant into fertilizer production. Congress insisted that Ford's five million dollar purchase price was too low. Progressive Republicans and Democrats united to block the sale. Harding's 1920 election sweep had not entirely healed the political scars of 1912.

During the 1920 presidential campaign, Warren Harding had insisted that government spending bordered on profligacy; that taxes were confiscatory; that government revenues fell short of outlays; that the national debt approached 25 billion dollars; that government intrusion into business operations appeared unremitting. The President sought less government in business as he promoted policies to stimulate growth, generate output, encourage capital investment, and enhance job opportunity. "Normalcy" represented an era before Wilson's labor government.

Andrew Mellon, Secretary of Treasury and Charles Dawes, Budget Director, spearheaded the president's drive for fiscal discipline. Dawes cut government spending and generated a budget surplus; Mellon rolled back excess corporate profits tax, cut estate taxes and began paying off government bonds.

Still, the transition from war to peace was not without an economic adjustment. In 1920-1921 the nation's output fell, retail sales collapsed, unemployment soared into double digits. The next year, paced by the growth of automobiles, radio broadcasting, telephone and telegraph, electric power, motor trucks, consumer products, and housing, the economy began to rebound.

In the 1920's radio was indelibly linked to the Jazz Age. In 1912, Congress set aside frequencies below 200 meter wave length for exclusive Navy use. Frequencies above long waves, defined as unproductive, were set aside for short wave amateurs.

The First World War disrupted radio amateurs use and application. After the war a patent gridlock reemerged, inhibiting radio production. Under government sponsorship General Electric formed RCA as a patent holding company and unblocked the patent freeze. U.S. radio. production took off. Frank Conrad, a wireless amateur in Pittsburg, Pennsylvania, began broadcasting radio signals in his local area. Others followed and the broadcast industry exploded. By 1923, some 200 firms had established radio stations in one form or another. Former corporate allies now found they were rivals. AT&T, the telephone company, began linking its radio stations via the company's long distance telephone lines. AT&T also prohibited non-Bell rivals access to the telephone companies toll network. In response, RCA attempted to create its own network through the use of Western Union's telegraph facilities. Telegraph lines, however, were ill-designed for voice transmission.[81]

GE and Westinghouse vied against AT&T for radio listeners. In 1926, AT&T executed an about-face, withdrew from its broadcasting venture, and sold its affiliate to RCA. The company named its new affiliate the National Broadcasting Company. Despite its withdrawal the Bell system had left a legacy to radio broadcasting. Radio commercials would be sold on the basis of time elapsed, not unlike the cost of a toll call. Commercial advertising revenues now fueled the growth of U.S. radio networks.

Congress in 1919, ordered the Navy to exit commercial radio activities. Internally, Harding's Post Office and Commerce Department jockeyed for regulatory authority over the new industry. It was here that the philosophy of Herbert Hoover and Harding coalesced. Although less than enamored by radio advertising. Hoover apposed government radio ownership. The industry, said the Commerce Secretary, "...would develop far more rapidly in this manner (competition) than if we pursued that European plan." In electing to retain private radio and telephone operations, the U.S. embarked on a policy quite apart from that experienced by Great Britain. In the latter half of the 19th century, Britain's Post Office licensed private telephone operations. The Post Office later acquired private systems and brought telephone services under Post Office ownership and control. [82]

The Post Office micro managed Britain's telephone network to such an extent that central management determined the location of individual telephone kiosks and the acquisition of motor scooters. Critics alleged that telephone employees tended to regarded themselves as tax collectors rather than purveyors of residential or business service. [83]

Britain's radio experience followed a similar pattern. The Post Office licensed a private firm, the BBC, to render radio broadcast service. By the mid 1920's, the Post Office assumed ownership control of the BBC and transformed it into a crown corporation.

Disdainful of advertising, the BBC banned commercials as a revenue source. Instead, the BBC funded its programs via a radio set tax. Listening to the radio, absent paying a tax, constituted a criminal offense. The BBC employed detection trucks to enforce its radio monopoly status. Hoover, Denby and Warren Harding obviously rejected not only the Wilson-Daniels radio-telephone model but Britain's Post Office ownership role as well. [84]

The Bell Telephone System (AT&T) and independent telephone companies invested in distribution, transmission lines and mechanized switching systems as local and long distance telephone to accommodate more subscribers. Not unlike the power industry, the telephone industry adopted a holding company to optimize their dispersed geographic affiliates.

Business and consumer usage fueled the nation's economic growth in the early 20s. The electric power industry, for example, invested more than one billion dollars in plant and equipment. In 1914, 30% of U.S. factories relied on electric power. By 1929, more than two-thirds of U.S. factories power would be supplied by central power turbines manufactured by General Electric, Allis-Chalmers and Westinghouse. [85]

Sales of fans, vacuum cleaners, refrigeration, ironing boards, air conditioning took off in the 1920's. U.S. passenger car production, paced by Ford, General Motors, the entry of Chrysler and a dozen independent car manufacturers, rose from 1.5 million in 1921 to 4.7 million by 1929. Albert Sloan of General Motors introduced the annual model changeover and encouraged installment credit buying. Sloan was able to balance centralized and decentralized corporate operations. Admiral William Moffett, BUAER, cited General Motors' corporate structure as a business model worthy of emulating by the navy. [86]

Automobiles, buses, trucks stimulated the nation's petroleum industry. In 1919, U.S. oil companies produced 378 million barrels of crude oil; ten years later the industry output had reached billion barrels annually. Searching to prevent preignition firing, General Motors and Standard Oil of New Jersey introduced a gasoline lead additive. Commercial airline industry would add to the demand for high octane aviation fuel.[87]

During the first World War, the U.S. government emerged as the largest buyer of aviation equipment. The fighting over, the government demobilized and an incipient aviation industry floundered as the Army and Navy dumped excess aircraft on the market. Harding supported Herbert Hoover's plan to develop commercial aviation through the introduction of landing lights, pilot safety requirements, and airport construction. And Denby offered to sell Navy flying boats at one third their costs as a device to promote private aviation. The Navy Secretary, of course, had a hidden agenda. In the event of war, commercial pilots would serve as naval reserve pilots. All told, President Harding wagered that a competitive industry could best ensure aviation's technical advance and development.[88]

The concept of commercial products embodying military application occasionally surfaced from unexpected sources in the 1920's. An employee of an organ and player piano company in New York, concerned by the expense of flying lessons, assembled a flight simulation machine in the basement of his father's shop. He would later trailer the unit to amusement parks and country fairs, charging 25 cents for a "pretend" airplane ride.[89]

The First World War disrupted world food production. U.S. agriculture prospered as a result. U.S. farmers secured bank loans and expanded acreage production in response to rising prices. After the war, however, European farming began to recover and commodity prices softened. Many U.S. farmers found themselves unable to service their bank debt, prompting the farm bloc to lobby congress promulgate policies reminiscent of the prosperous days of pre- 1914.

Anemic growth in the agriculture sector did not remove the fact that overall the economy prospered in the 1920's. Production increased, prices held stable, capital investment deepened, real wages trended upward. Industrial equipment and electrical suppliers, turbine manufacturers, boiler makers, machine tools, die and fixtures, the steel and metallurgy industry, electric power, radio, vacuum tubes, generators, gasoline motors, diesel engines, aviation—all experienced growth, abetted by an economic environment hospitable to investment, risk and capital expenditures.

MOBILIZATION

After Congress's war declaration in 1917, the U.S. struggled to jump-start an industrial mobilization program. As the demand for Army and Navy equipment exploded, the Wilson administration labored to balance arms production against an insatiable arms requirement. Unable to do so, government spending spilled into higher prices. The Wilson administration struggled for thirteen months before Bernard Baruch, Director of the War Industries Board, was given enough authority to convert U.S industry into war output and production.

In 1922, Secretary of War John Weeks and Navy Secretary Edwin Denby signed an executive order creating a Joint Army, Navy Munitions Board (ANMB). The Board took as its mission the coordination of material requirements of the two armed services, to forestall an Army-Navy bidding frenzy that had characterized the 1917-1918 mobilization era. [90]

At a secretary's council meeting Assistant Secretary, Theodore Roosevelt Jr., acknowledged that the Navy had fallen behind the Army in addressing the issue of mobilization planning. Theodore Roosevelt, Jr., distressed by the Navy's lack of progress, asked the Bureau Chiefs to assign a Navy officer to the ANMB. Rear Admiral C.B. McVay, Chief of the Bureau of Ordnance, said that his officers were "working pretty hard," implying that few officers could be spared for an ANMB assignment. Rear Admiral Coontz volunteered a War Plans officer to fill the ANMB slot. At the time, the navy's budget stood over 300 million dollars; the Chiefs of Naval Operations budget slightly under $30,000. Although the CNO detailed an officer to the ANMB, the navy's mobilization planning was off to a less than auspicious start. [91]

During the Great War, the Army expanded five-hundred times; the Navy ten times. Given its shore establishment of yards and arsenals it was perhaps natural that some bureaus viewed their in-house plant as insurance against the material needs of a national emergency. The Bureau Chiefs, however, went further. They challenged both the legitimacy of the ANMB and the authority of the Assistant Secretary of the Navy. The material bureaus defined mobilization as essentially revving up the department's shore establishment, leaving any spillover to commercial ship yards scattered along the U.S. coasts. The Bureaus, in short, could well take care of their own procurement requirements.

Nevertheless, the ANMB commenced a series of joint committee's meetings as an armed force coordination system. By October 7, 1922, the joint committee began looking at procurement policy, general supply, ordnance, medical, communication appliances, chemical warfare, power plant, machinery and tools, marine construction. [92]

INTER-GOVERNMENT COORDINATION

In 1913, despite the fact that the Joint Board constituted the venue for effecting inter-service supply coordination, President Wilson had taken the Joint Army-Navy Board (joint board) to task when the Board attempted to outline possible war options. On April 18, 1923, Denby issued General Order 103, a measure designed to improve liaison between Naval Districts and Army Corps areas. The order stipulated that both services participate in a monthly joint planning committee. The agenda ranged from shared facilities to shared munitions. Two months later the Navy delivered 2,000 excess 16 inch its projectiles to the U.S. Army. [93]

The Navy's Control Force was designated to assist the U.S. Army as well as U.S. Marines in any amphibious endeavor. Indeed, the 1923-24 Caribbean exercise would constitute the first such joint operation since the First World War.

In 1923 President Harding released an executive reorganization plan that proposed to fold the War and the Navy Department into a single entity—a department of defense. The proposal also anticipated that the nation's private sector would play a critical role in any future industrial mobilization. By the time Congress commenced hearings on the defense department proposal, Harding had died and Calvin Coolidge was now President. [94]

In a final move the Army-Navy Joint Board invited the State Department to attend its military planning sessions. On more than one occasion, the Navy turned to the State Department for advice on foreign policy matters. Denby, as noted, cleared the transfer of the Battle Fleet to the Pacific with Secretary Hughes. Hughes sent the request to President Harding who approved the transfer but suggested the announcement be delayed until the disarmament negotiations were completed.

In the early 20's Costa Rica and Panama squared off in a national boundary dispute. The State Department debated whether the Navy should dispatch Marines or warships to the area. These and other occasions prompted Edwin Denby to observe that, in times of peace, he often spent more time dealing with diplomats than with the armed services. [95]

The Army-Navy Joint Board concluded that its war plans required attendance of the Secretary of State. In late 1921, Edwin Denby and John Weeks invited Secretary Charles Evans Hughes to attend the Join Army-Navy Board session. Hughes replied that such attendance would be "impractical." The Board tendered a second invitation and suggested that, in lieu of the secretary, a State Department representative would be welcomed. Hughes left the possibility open. Interdepartmental coordination was apparently ahead of its time. A gap persisted nevertheless, leaving open the possibility that civilian policy might fritter away any military achievement. Weeks and Denby, both understood that military success stood as a means, not an end in and of itself. [96]

In sum, Warrant Harding viewed national security as dependent upon the vitality of the U.S. economy. The President fostered an environmental receptive to capital formation, income growth and employment. The President sought to impose discipline over government spending—hence a Bureau of the Budget. The naval arms limitation agreement marked one step in liquidating the cost of the nation's war effort. Finally, the President supported an economic environment congenial to expanding the nation's infrastructure, policies that accommodated electrical power, telephones, roads, and airways, wireless, oil pipelines. Once government spending was under control, federal revenues enabled the government to pay down the nation's war debt, reduced interest charges, and pass the savings to the nation's taxpayers.

In a post war environment of industrial change Denby's administrative task was equally challenging. In the 1920's U.S. corporations sought to strike a balance between decentralized and centralized operations—an optimum achieved by Alfred Sloan of General Motors. Sloan had inherited a random assembly of automobile operations, each marching to their own financial drummer. Financial decentralization almost destroyed General Motors.

To achieve a modicum of department rationalization, Denby enlarged the scope of his own office and increased the coordinating authority of the Office of the Chief of Naval Operations. He appointed the CNO as the department's budget officer; accorded War Plans cognizance over bureau's ship repair and alteration; endowed the CNO with fleet training and annual exercises; and attempted to dampen the centripetal force of the Navy's Bureau system. There the comparison between Edwin Denby and Alfred Sloan would end. Sloan was accountable to GM's shareholders; Denby was beholden to the House and Senate Naval Affairs Committee, a Congress that had decreed that four federal executive agencies should regulate Alaskan bears. "…two in the Department of Commerce, one in the Department of War and one in Treasury." Denby and Sloan resided in two alien worlds.[97]

NOTES

1. Paul R. Leach, *That Man Dawes*, (Chicago: Reilly and Lee, 1930), 175.

2. Leach, 193.

3. Charles R. Dawes, *The First Year of the Budget of the United States*, (New York: Harper, 1923), 60, 69.

4. See also Warren G. Harding papers. Letter from Warren G. Harding to Walter F. Brown, Executive Reorganization, June 11, 1921, Dartmouth College Library; U.S. Navy, *Report of the Secretary of the Navy, 1922*, 26; also U.S. Congress, House of Representatives, Information Requested of the Secretary of the Navy Under H. Res. 204, Edwin Denby to Thomas S. Butler, Naval Affairs Committee, 68th Cong.125.Sess. March7, 1924; see also Warren G. Harding papers, Letter from Warren G. Harding to Walter Brown, Senate office Building, June 11,1921, Roll #143, Dartmouth College Library, Harding sought to group "…each department shall be made up of agencies having substantially the same major purposes…"

5. Edwin Denby, Denby papers, Burton Historical Library, Detroit Public Library, Box 5, "History of the Naval Petroleum Reserves as shown by the record of the navy department," 16 May 1922, 18 Denby's letter to President Harding was delivered to the White House by Assistant Secretary, Theodore Roosevelt Jr. the Secretary's signature was unintentionally left out together with Rear Admiral Robert Griffin opposition.

6. U.S. Navy, *Annual Report of the Navy Department*, (Washington: Government Printing Office, 1921): 6.

7. NARA, RG 80, *Secretary's Council*, July 14, 1923, 70. Acting secretary Roosevelt said, "If we send a battleship to a yard, say she has 200 items to have done, a raft of them under engineering and a raft under construction and repair etc., each item carries a separate job order." Roosevelt wanted to reduce the paper work to one job order per battleship.

8. Congressional information service: *CIS index to U.S. Executive Documents, 1920-1932*, (Behesda: CIS 2001) 10 General Order #53, cited as CIS Index.

9. Gerald E. Wheeler, *Admiral William Veazie Pratt, U.S. Navy: A Sailor's Life*, (Washington: Department of the Navy, 1974), 180.

10. NARA, RG80, Secretary's Minutes, January 31, 1923, 8.

11. CIS Index, General Order No. 84, Regulations Governing the Operation of the Experiment and Research Laboratory, 25 March, 1922.

12. U.S. Naval Administration in World War II, Administrative History, History of Naval Research Laboratory, Vol. 134, Exhibit 2. General Order No. 84, 25 March 1922, "Regulations Governing the Operation of the Experiment and Research Laboratory," Edwin Denby.

13. A. Hoyt Taylor, *The First Twenty-Five Years of the Naval Research Laboratory*, (Washington: U.S. Navy Department, 1948); 17 [radio detection of a moving vessel].

14. Admiral Harry Hill, Columbia University Oral History, Research Office, No. 10, Columbia University, November 6, 1967, 72.

15. Theodore Roosevelt, Jr., *Diary*, June 13, 1922, 310, Manuscript Division, Library of Congress.

16. NARA, RG 80, Memorandum from Director War Plans to Chief of Naval Operations, 8 October, 1921; also Edward S. Miller, *War Plan Orange: The U.S. Strategy to Defeat Japan, 1897-1945*, (Annapolis: Naval Institute Press, 1991): 115.

17. Naval Administration, Selected documents on Navy Department Organization: 1915-1940, U.S. Naval War College Library, 1945, IV-5, (cited as selective navy documents).

18. Warren G. Harding, Warren Harding papers, memorandum for the Director, Bureau of the Budget, October 24, 1921, Roll 141, Dartmouth College, (cited as Harding papers).

19. U.S. Navy, Annual Report of the Navy Department, Bureau of Engineering, 1924, 294.

20. NARA, RG 80, Minutes of The Secretary's Council, July 14, 1923.

21. Congressional Information Service: *CIS Index to U.S. Executive Documents, 1920-1932*, (Washington: Government Printing Office, 1922) General Order No. 103. Establishment of an office of fleet training, 6 June 1923, (Cited as CIS index).

22. U.S. Navy, *Report of the Secretary of the Navy*, Report of Chief of Naval Operations, 1924, 79, "two officers [Division of Fleet Training] attended the joint maneuvers of 1923-24 as umpires and four officers assisted in a month's committee work at the Army War College.

23. U.S. Navy, Bureau of Ships, *An Administrative History of the Bureau of Ships During World War II*, Vol. 3, 1952, 73; *Marine Engineering and Shipping Age*, (May 1922): 280; U.S. Navy, *Report of the Secretary of the Navy*, 1923, 8-9.

24. Ronald Spector, *At War At Sea: Sailors and Naval Combat in the Twentieth Century*, (London: Penguin, 2001), 137.

25. William F. Trimble, *Admiral William Moffett: Architect of Naval Aviation*, (Washington: Smithsonian Institution Press, 1994), 136.

26. *Army and Navy Journal*, (February 24, 1923) "It has been decided to transfer the management of the reserve ship from the bureau of navigation to naval operations..."

27. CIS Index, 10; also NARA, RG80. Office of the Chief of Naval Operations and Office of the Secretary of the Navy, 1919. 1917, Memorandum from the CNO to Commander of the Fleet, Organization of the Fleet, July 8, 1921.

28. Warren Harding papers, Bureau of the Budget, Special advisor, "The Steamship Great Northern," October 25, 1921, #41, Dartmouth College Library. "In my judgment the reasoning of the Navy Department as to the necessity is correct and in accordance with sound military tactics," see also, Warren G. Harding papers, memorandum for the director bureau of the budget, October 24, 1941, Roll #141, Dartmouth College Library.

29. Wheeler, 191.

30. Roger Dingman, *Statesmen, Admirals and Salt: The United States and the Washington Conference, 1921-1922*, (Santa Monica: California Arms and Foreign Policy Seminar, 1973), 11. Also Wheeler, 180.

31. Lawrence Madaras, "The Public Career of Theodore Roosevelt, Jr." Ph.D. dissertation, New York University, 1964, 180.

32. Edwin Denby papers, *United States vs. Mammoth Oil Company*, in the District Court of Wyoming, deposition of Rear Admiral John K. Robison, July 21, 1924, Box 8, 70.

33. Lawrence H. Douglas, "Robert Edward Coontz, 1919-1923," Robert William Love, Jr., ed., *The Chiefs of Naval Operations*, (Annapolis: Naval Institute Press, 1980), 32.

34. U.S. Congress, House of Representatives, Air Services Unification, Supplemental Naval Appropriations Bill, 11922, testimony of General William Mitchell, Feb. 1, 1921, 61st Congress.

35. U.S. Congress, House of Representatives, Committee on Naval Affairs, testimony of Assistant Secretary Theodore Roosevelt, Jr., 67.

36. John J. Reber, "Pete Ellis: Amphibious War Prophet," Merrill L. Bartlett, ed., *Assault from the Sea: Essays on the History of Naval Warfare*, (Annapolis: Naval Institute Press, 1983), 158; also NARA, RG 127, E.H. Ellis, Naval Bases: Location, Resources, Denial of Bases, Security of Advanced Bases: U.S. Marine Corps, Division of Operations and Training, Historical Amphibious File, Marine Corps Education Center, Quantico, Virginia, 1921.

37. John A. Lejeune, *The Reminiscence of a Marine*, (Philadelphia: P.A. Dorrance, 1930), 474-475.

38. E.B. Potter, *Nimitz*, (Annapolis: Naval Institute Press, 1976), 138-140.

39. Edward S. Miller, War Plan Orange: *The U.S. Strategy to Defeat Japan, 1897-1945*, (Annapolis: Naval Institute Press, 1991), 79.

40. 79.

41. NARA, RG 80, Record of General Board, Major General Commandant Memorandum for the General Board, *Future Policy of the Marine Corps*, February 11, 1922, File 431.

42. NARA, RG 80, Secret and Confidential Correspondence, Letter to Senator Miles Poindexter, from Edwin Denby, October 15, 1921, 72.

43. Arthur E. Burns, III, *The Origen and Development of the U.S. Marine Corps Tank Units: 1923-1945*, (Quantico: Marine Corps Command and Staff College, 1977.

44. NARA, RG 80, Hearings Before the General Board, U.S. Naval War College Library, June 26, 1922, p. 569-570.

45. U.S. Congress, House of Representatives, Hearings, Authorizing the acquisition of certain sites for naval air stations," H.R. 11983, June 8, 1922: also H.R. 10274, Marine corps Flying Field, January 24, 1922, H.R. 10274, House of Representative 67, Congress, Sess. 3, 1148-1149.

46. NARA, RG 127, U.S. Marine Corps, Division of Plans and Policies War Plans Sections 1915-1946, Box 2. General Correspondence pertaining to war portfolio, memorandum from secretary of the navy to war portfolio. Subject: Control Force, United States Fleet, 15 August, 1921, Breckinridge Library, Quantico, VA.

47. Holland M. Smith, *Coral and Brass*, (New York: Charles Scribner & Sons, 1949): 19.

48. Harding papers, "Memorandum for the President from Edwin Denby," April 19, 1921, Roll #141, Dartmouth College Library.

49. John A. Lejeune, "Amphibious Landing," Lecture, U.S. Naval War College, June 1924, Naval War Collection, U.S. Naval War College. See also NARA, RG 127, Records of the U.S. Marine Corps Division Plans and Policy, War Plan Section, 1915-1946, Box 2, memorandum for the Major General Commandant, from P. Halford, 3 August, 1923, Breckinridge Library, Quantico, VA.

50. Ralph Earle, "Landing Operations of the Control Force," Lecture, U.S. Naval War College, 11 December 1922.

51. United States Shipping Board, Fifth Annual Report, (Washington: Government Printing Office, 1921), 42-43; also Sixth Annual Report, (Washington: Government Printing Office, 1922), 179-180. Also David H. Stratton, *Tempest Over Teapot Dome: The Story of Albert Fall*, (Norman: University of Oklahoma Press, 1998), 245-246. See *New York Times*, February, 15, 1920, 10.

52. Edwin Denby, Burton Historical Collection, Detroit Public Library, Admiral Robison's (Robison's) deposition, July 25, 1924, Box 5, 10.

53. David H. Stratton, *Tempest over Teapot Dome: The Story of Albert Fall*, (Norman: University of Oklahoma Press, 1998), 245-246. See also United States Reports, 275, The Supreme Court, *Mammoth Oil Co. v. United States*, October, 1927, 36. The Salt Creek Field was "very productive." Also U.S. Congress, Senate, Leases Upon Naval Oil Reserves, rear Admiral John K. Robinson, November 20, 1923, 68 Congress, Sess. 2nd; see United States Shipping Board, Sixth Annual Report, June 30, 1922, p. 179, also *New York Times*, February 15, 1920, 10; U.S. Congress, Senate, Committee on Naval Affairs, History of Naval Petroleum Reserves, 78th Cong. Session 2nd, April 12, 1944, p. 21.
The Walsh committee had retained its own geologists to report on the surface conditions in Teapot Dome. Their conclusions were that Teapot Dome and the neighboring Salt Creek were parts of the same geological structure and that drainage could take place as to a portion of the oil in the reserve.
Warren G. Harding papers, memorandum from secretary of the navy to commander-in-chief, Atlantic Fleet, August 10, 1921, 309, Roll #141, Dartmouth College Library.
54. Weldon Winans Harris II, "Harry F. Sinclair and the Teapot Dome Scandal: Appearance and Realities," master's thesis, Indiana University, June 1961, p. 66. P.C. Spencer, *Oil and Independence: The Story of the Sinclair Oil Corporation*, (New York: Newcomen Society of North America, 1957), 19, "Of the 135 to 170 million barrel of oil predicted for it, it produced less than two million," See also Robert L., Owen, *Remarkable Experiences of H.F. Sinclair with his Government* (Tulsa: H.H. Rogers, 1929), 8.
55. NARA, RG 80, Confidential Correspondence, memorandum from Edwin Denby to John Weeks, Secretary of War, Panama Canal, December 5, 1923.
56. NARA, RG 8, Confidential Correspondence, Secretary's Council, "All possible work to maintain the above described force will be thrown to the navy operation's stations, Pearl Harbor, both in the line of ships repairs and of manufacturing work," June 19, 1922, 103. See also CINCUS, Annual Report, 27 September, 1927, p. 13 "Lesson learned from each concentration since 1922 have resulted in carrying additional equipment so that forces afloat have become less dependent on shore establishments.
57. NARA, RG 80, Confidential correspondence, memorandum from Chief of Naval Operations to Board in the Development of Navy Yard Plans, Mobile Base, June 20, 1922, 2.
58. Warren G. Harding, Warren Harding papers, memorandum from Warren Harding to Edwin Denby, October 20, 1921, Roll #176, 406; also, memorandum from the Bureau of Engineering to the Commander, United States Fleet regarding spare parts, 3 January 1922, Roll #141, 406, Dartmouth College Library, Hanover, New Hampshire.
59. Denby, Edwin. Denby papers, Bentley Historical Library, University of Michigan, Box 2, Draft, Secretary of the Navy to U.S. Senate, Naval Appropriations, 28 April, 1922, 3. "...I issued on August 25, 1921 an order ... Limiting repairs to ships to those necessary for military efficiency..." "The present General Order further limits repairs and alterations, in as much as it limits officers afloat to use of repair ships and the elimination of all possible work at shore stations where the work is more expensive." General Order No. 67. "it is directed that commanders in chief and subordinate commanders use every effort to effect routine maintenance and up keep of their ships by ships' forces with the facilities on board, plus those of the repair ships, i.e. the fleet maintain the fleet to the greatest degree possible." Also Harding papers, from Chief of Naval Operations to Commander in Chief, Pacific Fleet, August, 1921. Roll #141, 313-314. "...the repair work to be undertaken by the yards must be confined to that which is beyond the capacity of ships' forces assisted by the repair vessels."
60. NARA, RG 80, confidential correspondence, memorandum from assistant director for material to director of plans division, seven tankers, U.S. Shipping Board. Assistant Director wants to acquire shipping board tankers at earliest possible date as floating storage. Sixteen tons of floating storage "...represents an increase in the offensive power of the navy." June 21, 1921. Warren G. Harding papers, memorandum from Warren G. Harding to Edwin Denby, transfer of shipping board tanker to the navy, October 21, 1921, 406, Roll 176, Dartmouth College Library.
61. NARA, RG 80, Secretary's Council, August 24, 1921, 148. Also, Frederick B. Laidlaw, "The History of the Prevention of Fouling," U.S. Naval Institute Proceedings, 78, 7 (July, 1952): 777. "In 1922, at the request of the Navy Department, experiments in hot plastic

antifouling paints were begun again by the chemical warfare service at the Edgewood Arsenal." Also, "…an extensive investigation of the entire problem of fouling was begun in September 1922, under the Bureau of Construction and Repair." Also, U.S. Smithsonian Institute Archives, J. Paul Fisscher papers, letter from H.R. Vickery, U.S. Navy, Boston to J.P. Fisscher, 3 March, 1925.

62. U.S. Congress, House of Representatives, Naval Affairs Committee, Hearings on Appropriations, Fiscal 1923, testimony of Admiral Robert Coontz, June 19, 1922, 411.

63. NARA, RG 38, Office of the Chief of Naval Operations, Fleet Training Division, General Correspondence, 1914-1941, Box 101, 443 (1-10_, 12-22-21, Rules for Engineering Performance, merchant marine vessels of the United States Merchant Marine.

64. Edwin Denby Paper, Bentley Historical Collection, University of Michigan, Draft Remarks delivered to secretary's office, October, 1923.

65. *Army and Navy Journal*, 60, October 28, 1922, 2034.

66. Harding papers, letter from Herbert Hoover, May 15, 1922, Roll 131, 869, Dartmouth College Library.

67. Indelible Mark on Advertising left by Lasker Agency Pioneer, *Advertising Age*, (June 9, 1952). Also Jeffrey L. Cruikshank, Arthur W. Schultz, *The Man Who Sold America*, (Boston: Harvard Business Review Press, 2010), 204.

68. Warren G. Harding, Harding papers, Dartmouth College Library, Albert Lasker, William Benson, latter to President Harding, May 8, 1923, 403.

69. Warren G. Harding, Harding papers, letter from Warren Harding to B.T. Meredith, Editor, Successful Farming, Dartmouth College Library, Roll #176, June 20, 1922, 1431.

70. The Congressional Digest, "Labor Organization Discusses Ship Subsidy Bill," No. 10, July 1922, 13.

71. Robert K. Murray, *The Harding Era: Warren Harding and His Administration*, (Minneapolis: University of Minnesota Press, 1969), 288.

72. U.S. Congress, Senate, To Amend Merchant Marine Act of 1920, 66th Cong., 2nd Sess. Testimony of Edwin Denby, May 16, 1922, 2189.

73. U.S. Library of Congress Manuscripts Division, *Diary,* Theodore Roosevelt, Jr., "Lasker turned up in the late afternoon to tell me that the Carolina senators had been to him and advised him that they would trade their votes for the ship subsidy if he, Lasker, would set the Charleston navy yard opened up." 340.

74. Warren G. Harding, Harding papers, letter from Warren G. Harding to Senator Reed Smoot, Dartmouth College Library, Roll #230, December 4, 1922, 80.

75. NARA, RG 80, confidential correspondence, Secretary's Council, June 6, 1921, 74-75.

76. NARA, RG 80, Secretary's Council, May 15, 1923, 42.

77. Gary E. Weir, *Building American Submarines, 1914-1940*, (Washington: Department of the Navy, 1991), 55.

78. NARA, RG 80, confidential correspondence, Secretary's Council, January 11, 1923, 9.

79. NARA, RG 80, confidential correspondence, Secretary's Council, September 8, 1921, 131.

80. NARA, RG 80, confidential correspondence, July 14, 1923, 57.

81. George H. Douglas, *The Early Days of Radio Broadcasting*, (London: McFarland, 1987), 73.

82. Eugene E. Wilson, *Slipstream: The Autobiography of an Air Craftsman*, (New York: Whittlesey House, 1950), 67.

83. Viscount Wolmer, *Post Office Reform: Its Importance and Practicality*, (London: Ivor Nicholson and Watson, 1932), also Alan Clinton, *Post Office Workers; A Trade and Social History*, (London: George Allen and Unwin, 1984), 28, on telecommunications, "The UPW (United Post Office workers) and others pressed the case for nationalization," Post Office archives, London, U.K; Post Master General, The Bridgeman Report, (committee of inquiry on the Post Office, 1932), Post Office archives, London U.K., "…we feel too much centralization, too little freedom left to the local office in the provinces,", p. 16, John Cadman.

84. Christopher H. Sterling, *The Museum of Broadcast – Communications, Encyclopedia of Radio*, Vol. 1, (New York: Fitzroy Dearon, 2000), Tim Crook, British Broadcasting Corpora-

tion," 203; R.H. Coase, *British Broadcasting: A Study in Monopoly*, (London: Longman Green, 1950), 86.

85. Samuel E. Morrison, Henry Steele Commanger, William Leuchtenburg, *The Growth of the American Economy*, (New York: Oxford University Press, 1980), 424.

86. Eugene E. Wilson, *Slipstream: The Autobiography of an Air Craftsman*, (New York: Whittlesey House, 1950), 67.

87. Harold V. Faulkner, *American Economic History*, (New York: Harper and Row, 1976), 342.

88. Harding papers, Edwin Denby to George Christian, September 13, 1921, Roll 230, Dartmouth College Library.

89. Susan Van Hoek, Marion Layton Link, *From Sky to Sea: The Story of Edwin A. Link*, (Flagstaff: Best, 1983), 20-21.

90. Robert H. Connery, *The Navy and the Industrial Mobilization in World War II*, (Princeton: Princeton University Press, 1951), 35. "Army and Navy Munitions board was created upon recommendation of Join Army and Navy Board (letter No. 346, serial No. 181, dated June 27, 1922...."

91. NARA, RG 80, Secretary's Council, January 31, 1923, 34.

92. U.S. Navy, United States Naval Administration, Office of the Chief of Naval Operations, procurement of Organization under the Chief of Naval Operations. *The History of Op-24*, Vol. 22, 1946.

93. U.S. Naval Administration in World War II, Procurement organization under the Chief of Naval Operations, *History of Op-24*. U.S. Navy Department Library, Washington, D.C. 13.

94. *Army and Navy Journal*, 60, 1 (May 19, 1923): 1. Also NARA, RG 80 Secretary's Council, June 31, 1923.

95. U.S. Congress, Joint Committee on the Reorganization of Government Departments, Reorganization of Executive Department, 67th Cong., 4th Sess., February 19, 1923.

96. William Young Smith, "The Search for National Security Planning Machinery," Ph.D. dissertation, Harvard University, 1960, 247.

97. NARA, RG 80, Joint Army-Navy Board, 301, Box 1, Joint Board Planning Committee, October 18, 1921.

Chapter Three

The Coolidge-Hoover Administration, 1923-1933

ADMINISTRATIVE OVERSIGHT

The Coolidge and Hoover administrations covered a period from August 1923 to March 1933, just short of a decade. The era was one of general prosperity followed by the 1929 stock market crash and the onset of the Great Depression. In the 1920's, the Bureau of the Budget settled in as the President's management tool. The Bureau scrubbed the Navy's proposed spending plans before submitting them to Congressional Appropriations and Naval Affairs Committees. In the decade of the 20's Navy annual budgets averaged between $250 to $300 million dollars.

Three secretaries served in office during this time period: Edwin Denby and Curtis P. Wilbur, under Coolidge; Charles F. Adams, Secretary under Herbert Hoover. From August 1923 to March 10, 1924, Denby was swept into the navy's oil reserve controversy. By February 1924 Denby tendered his resignation to President Coolidge.

Curtis Wilbur, Denby's successor, an attorney, had also graduated from the U.S. Naval Academy. In 1929, Herbert Hoover succeeded Coolidge as President and chose Charles F. Adams as Wilbur's successor. Several officers presided as Naval operations head including Admirals Edward Eberle, Charles F. Hughes, and William V. Pratt.

In 1927, Rear Admiral Thomas P. McGruder, Commandant of the Philadelphia Navy yard, wrote an article in the *Saturday Evening Post*. McGruder charged that Congress, failing to develop Pearl Harbor's defenses, had instead placed a drydock at Charleston, South Carolina. The admiral's article triggered a firestorm. Secretary Wilbur relieved McGruder from his Navy Yard post. He remained unassigned for 18 months, then was detailed to the

Pacific Base Force, pending retirement. In case they forgot, the navy's officer corps were reminded, once again, of the power and influence of congressional naval affairs committees. [1]

Charles Adams, Hoover's Navy Secretary, had previously served as Harvard University's chief financial officer. During Adams' tenure in office, the
economy collapsed, unemployment rose, and falling government revenues
squeezed naval budgets.

In 1930, the Navy transferred the Naval Research Laboratory from the
Secretary's office to the Bureau of Engineering under Rear Admiral Samuel
M. Robinson. Commander Stanford C. Hooper, Director of Naval Communications, reported to Admiral Robinson. Sometimes referred to as the "father
of naval radio," Hooper was wedded to continuous long wave technology. He
thus took exception to the research direction of the laboratory. According to
Ivan Amato, Hooper favored "...abolishing the laboratory," or at least confining the laboratory to testing radio apparatus. [2]

In 1930 Lawrence Hyland, a laboratory technician, detected a radio echo
from a passing plane. Hyland repeated his experiment several times to confirm the discovery. Dr. A. Hoyt Taylor, NRL, wrote a memo to the navy's
upper echelon implying a need for additional funding. The bureau's response
was perfunctory. Actually, the Navy thought the radio detention phenomenon
might be more appropriate for the Army. [3]

Chief of Naval Operations (CNO)

Denby, meanwhile, continued to upgrade the authority of Operations. His
appointed of Admiral Robert Coontz as the department's budget officer ignited a backlash from Representative James F. Byrnes, however. Byrnes noted
that the Secretary had given the CNO responsibility over warship repair and
alteration. Byrnes argued that Denby's policies had reduced the role and
status of the Navy Secretary. According to Byrnes, a new Secretary had
better bring his golf sticks to Washington because his work had been pre-
empted by the navy's ranking admiral. [4]

To arrest what he perceived as an erosion of civilian authority, Byrnes
offered an amendment to a forthcoming appropriations bill that,

> Provided that no money appropriated by this act shall be available for the pay
> of any commissioned officers of the Navy while attached to the Office of
> Naval Operations work not specifically assigned by law to such office. [5]

The amendment passed the House of Representatives, but failed in the Senate. Although the amendment remained stillborn, the message once again
served as reminder to the Navy's officer corps. Congress would countenance
little interference with its budget and bureau authority.

In 1930, Admiral William Pratt succeeded Admiral Charles F. Hughes as Chief of the Office of Naval Operations. Pratt had voiced reservations over the Navy's command structure as well as the fleet's Control Force. Now CNO, Pratt was in the position to do something about it. He oriented the fleet back toward a type command format and designated the Control Force as the submarine force. It was also during this period that the Marine Expeditionary Force was assigned to the Navy's Special Service Squadron, sometimes referred to as the State Department's Navy.

BALANCED FLEET

During the 20's the Navy continued to upgrade its existing battleships from coal to oil fired boilers. At the same time the Navy reduced the weight of its heavy warships, although the department added horizontal armor as protection against possible air attack.

One decision put Denby in Coolidge's dog house. Under the Washington agreement, navy's gun turrets had been restricted to an elevation of 18%. The General Board and the Bureau of Ordnance informed Denby that Britain's guns were capable of 25% elevation, placing U.S. warships at a disadvantage. Denby requested and received $6.5 million dollars from Congress to match the Royal Navy's modernization program.[6]

Denying the gun elevation charge, the British Admiralty, insisted that they had adhered to the letter and spirit of the naval limitation agreement. Misinformed, Denby had to return the $6.5 million dollars to the Treasury Department. On February 18, 1924, President Coolidge penned Denby a cryptic note: "I wish you would confer with me about the proposal to raise the elevation of some of the guns on the capital ships, before you take any action on it."[7]

By now carrier *Langley* had joined the U.S. fleet. A thunderstorm in Ohio broke the calm in 1924. The *Shenandoah*, a Navy airship crashed, killing several crew members. Now assigned to San Antonio, Texas, Colonel Billy Mitchell charged that those responsible for the airship's destruction had committed an act that bordered on treason. Mitchell renewed his belief that the long distant bomber had consigned the battleship to the dustbin of history.

Mitchell's allegation prompted Secretary Wilbur to appoint a board to examine the relative merits of battleship and carrier aviation. Chaired by Admiral Edward Eberle, some eighty witnesses appeared before the board. A few officers testified that, given armor improvements, blisters, anti-aircraft guns, fire-control techniques, propulsion machinery, the future of the battleships remained secure. Rear Admiral William Fullam took exception to that scenario. As Commander of the Pacific battle fleet in 1918, Fullam had witnessed 130 planes flying over his fleet in a post war celebration. At that moment, Admiral Fullam concluded that naval aviation had tilted the tactical scale against the heavy warship. [8]

The Eberle final report concluded that aviation's future was limited by the laws of physics. The battleship, by contrast, had yet to reach its technological potential. The board granted that although aviation was essential to the fleet, the battleship remained the backbone of the fleet. [9]

Calvin Coolidge weighed into the aviation controversy. He appointed Dwight Morrow, a former member of J.P. Morgan Company, as head the President's Aviation Board. The Morrow Committee solicited the views of U.S. military officers as well as executives from the nation's aviation industry. Colonel William Mitchell, testifying for two days with little interruption, insisted that the U.S. adopt an independent air force (a merger of army and Navy) and that military and private aviation be placed under the authority of a single government agency.

The Morrow report, brushing Mitchell's proposals aside, insisted that each military service retain, operate, and control their individual aviation unit. The committee proposed that each service add an Assistant Secretary for Air and recommended that the U.S. Army Air Service be renamed the U.S. Army Air Corps. Finally, the board proposed that Congress fund military aircraft production for the next five years.

Although the tonnage limitation on battleships and carriers remained in place throughout the 1920's the Washington Treaty did not end naval rivalry among former allies. Instead, Britain, Japan and the U.S. shifted to cruiser, destroyer and submarine construction, the unregulated side of the combatant equation. The Coolidge administration sent representatives to Geneva, Switzerland to bring non-regulated vessels under tonnage constraint. Influenced by the absence of fixed bases in the western Pacific, the U.S. Navy sought 10,000 ton cruisers endowed with ample fuel capacity. Britain, given its network of global bases, insisted that 6,000 ton cruisers mounting six inch guns, constituted a more appropriate standard. Negotiations deadlocked and U.S. representatives came home empty handed. President Coolidge and Congress proceeded to raise the U.S. bargaining ante by authorizing the construction of 15 cruisers. [10]

Herbert Hoover, entering the White House in March 1929, continued Coolidge's effort to bring all warship classes under some tonnage regulation. Convened in London, the parties set cruiser assignments at 6,000 tons bearing 6-inch guns. The party's accepted 10:10:09 tonnage ratio to the U.S., Britain, and Japan respectively. The London Agreement placed a ceiling destroyers and submarines as well. All parties, needless to say, came away dissatisfied. U.S. Representatives had sought heavy cruisers, Japan had wanted naval parity. One Japanese naval officer, in fact, vented his frustration by assassinating the country's Prime Minister.

The Washington (1922) and London (1930) treaties now blanketed all combatant classes. The coverage did not resolve a domestic political issue. Would Congress appropriate funds to bring the fleet to its assigned treaty level? The answer remained uncertain. The 71st Congress continued its parsimonious mood, and with the onset of the depression, government revenues experienced a free fall that made naval funding more unlikely. Congress and the administration cut federal spending across the board and proceeded to raise taxes in an effort to balance the government's budget.

Throughout the 1920's, submariners remained frustrated by the limited range of their boats. During annual fleet exercises, submarine diesels experienced countless breakdowns. Given the absence of a naval base west of Hawaii, submariners sought a power plant capable of a 10,000 mile range and a surface speed of 20 knots. Submarine officers were so dissatisfied with their in-house diesels that they sent Admiral Harry J. Yarnell to Europe to explore the possibility of licensing an engine from Germany. [11]

Fleet Exercises

During the 1920's, fleet exercises continued as an annual event. At the end of each fleet problem, naval officers submitted written comments, observations and recommendations to CINCUS. The reports ranged from the status of carrier aviation to refueling ships at sea, from tactical training instructions to questionable umpire calls.

The fleet's annual maneuvers enabled the Navy to test gunnery, communications, engineering, and tactical coordination. Working with Operation's War Plan Division, the Division of Fleet Training condensed comments into instructional booklets, guidelines and manuals to be made available to officers and commanders alike. Nor were participating officers hesitant in commenting on the status of the publications. Concerned by so many additions and changes, one respondent suggested that a loose leaf binder might be more appropriate for officer edification. Others noted discrepancies between tactical guidelines, battle instruction and signal books. Annual fleet exercises, nevertheless, tended to integrate War Plans, Fleet Training, CINCUS, and on occasion, the material bureaus. Denby reminded the Navy that fleet readiness constituted their overriding priority. [12]

Throughout the decade, the battleship stood as the fleet's premier capital weapon, all other ships defined by their dreadnaught support—a ranking that influenced an officer's career path. Ensigns frequently requested battleship assignment as an essential step to their career advancement. The fleet's cruising formation reflected the centrality of the heavy warship. The battleship stood at a central ring, other combatant ships residing in outer concentric circles.

The *U.S.S. Langley* joined the fleet in 1925, participating in artillery spotting and fleet reconnaissance. Four years later, the *Lexington* and *Saratoga* joined fleet exercises in the Caribbean. As in previous problems the fleet was divided into apposing forces, one attacking the Panama Canal, the other defending it. The *Saratoga* was assigned to the offensive fleet. Accompanied by the cruiser *Omaha*, the *Saratoga* raced toward the Galapagos Islands at night, turned east and at five o'clock the next morning launched a 70 plane attack. The strike caught the Army and Navy Panama defense forces by surprise. [13]

Although aviation partisans celebrated the *Saratoga's* run, a fleet umpire ruled that a battleship had sunk the carrier. The *Saratoga* had, nevertheless, pulled off an unorthodox tactic, an exercise that precipitated a series of questions. Could battleships, cruisers and destroyers match the speed of the carrier? Was the carrier capable of offensive action? Were catapult planes still effective as fleet scouts?

The 1929 exercise also triggered an internal debate as to the very content of fleet tactics. Lieutenant Commander Eugene Wilson, Admiral William Pratt's chief of staff, had proposed the *Saratoga's* offensive strike. After the exercise, however, Wilson picked up signs that the *Saratoga's* unconventional run had somehow sullied his naval career. Shortly thereafter, Commander Wilson resigned from the service. [14]

Despite doctrinal debates between surface and aviation officers, the principals involved agreed that aviation was now integrated with the fleet. What remained unanswered was whether carriers should operate as an independent attack force or remain tied to the main fleet. [15]

In 1930, Rear Admiral Henry V. Butler suggested that cruisers and destroyers should accompany a carrier on future raids, noting that the *Saratoga's* night run "...is of the highest importance for study and exercises in the future." Rear Admiral W.C. Cole went further and proposed that a carrier task force include a division of four cruisers and eight destroyers. Commander Forrest Sherman published an article in U.S. Naval Institute Proceedings suggesting that 10,000 ton carriers might possess an offensive capability.

Admiral Harold Wiley (CINCUS) took exception to Admiral Pratt's unorthodox carrier run. Wiley insisted that in future engagements the battleship stood as the final arbiter of any fleet action. The Lexington Class carriers, however, would influence the design and architecture of future carrier generations.

Casting their eyes beyond the Navy's in-house diesels, submariners began to monitor developments in the U.S. commercial diesel market. In 1930, General Motors acquired the Winton Engine Company, a manufacturer of diesel engines. The merger signaled GM's entrance into the railroad propulsion market. In a matter of months, Winton and U.S. submariners would find themselves entering into exploratory discussions. [16]

The Navy's annual fleet problems repeated the usual set of gunnery, signally, radio communication, fuel consumption exercises. Crews and ships received either a trophy or money for outstanding performance. On occasion a torpedo malfunctioned or ran cold, prompting an umpire to deduct points against the offending ship and crew. In the fall of 1923, Rear Admiral S.S. Robison, Commander of the Navy's submarine force, routed a memo to the CNO via CINCUS. Admiral Robison proposed that torpedo exercise rules be modified so that a ship's crew not be penalized if one of its torpedoes malfunctioned. The Bureau of Ordnance replied to Rear Admiral Robison's memo: "Our torpedoes can be made to function properly only by experienced and careful torpedo persons. If you make a mistake, don't try to cover it up or present an alibi."

In 1923, submariners stood reprimanded. [17]

The 1923-24 winter maneuvers included a concentrated U.S. fleet under the direction of a single Commander in Chief, (CINCUS). Following the exercises, Admiral Robert Coontz recommended that arc transmitting radio sets be replaced by vacuum tube sets. He also proposed that upgraded radio equipment be extended to all ship classes. The Navy was beginning to define radio as essential to tactical coordination. [18]

AMPHIBIOUS OPERATIONS

By the mid twenties Major General John Lejeune had outlined the components of an expeditionary force—landing craft, transports, air support, ground artillery, cargo vessels, battleship support, logistics, tanks, radio equipment. Amphibious exercises were scheduled to take place in the Caribbean during the winter of 1923-24. In anticipation, the Marines requested an additional transport. On more than one occasion Edwin Denby reminded President Harding that, without transports, the Marines would be forced to employ battleships, a solution unsatisfactory to both the fleet and the Marines. In a funding request, Congress had excised a new transport ship from the Navy's construction list.

In any attack from the sea, the Marines anticipated that defending forces would employ artillery. The Marines sought a weapon in kind. Denby asked the Bureau of Construction and Repair to construct a lighter that accommodated a 150mm howitzer and its associated tractor. The Bureau responded that a towed barge appeared to be a workable answer. [19]

Landing exercises did provide a platform for Marine Corps experimentation. For one thing, standard Navy boats simply could not handle normal surf conditions; existing Navy transports were ill-suited for joint operations; radios, exposed to salt water were unreliable or unworkable; and night landing operations viewed as a prescription for chaos.

On a more promising note, the Marines insisted that battleship fire constituted an essential complement to any invasion endeavor. The Marines also
concluded that "combat loaded" transports expedited the flow of supplies to
units ashore. In addition, the 1923-1924 winter exercises enabled the Marines
to experiment with a British personnel landing craft as well as an amphibious
tank prototype. Although both were less than successful the exercises served
as a venue for equipment testing and experimentation.

In the fall of 1924, Admiral Robert Coontz, CINCUS, submitted an evaluation of the fleet's concentrated exercise. The Navy, wrote Coontz, should
complete the *Lexington* Class carriers, improve troop transport designs, upgrade radio sets, develop an amphibious command ship, and build the fleet to
a tonnage level permitted under the Washington arms agreement. [20]

By this time, Lejeune concluded that the Marines should incorporate their
own landing craft, air support, transport and cargo vessels, and field artillery.
In a lecture to the Navy's War College, Lejeune proposed that battleships be
detached temporarily from the fleet to give the Marines prelanding fire support. [21]

The next year, the Marines conducted exercises in Hawaii. Following that
maneuver, the Marines were assigned to China, Central America, and Nica-
ragua. In 1927, a Marine ground unit, surrounded by Nicaraguan rebels in
Octal, (Sandinistas), signaled for Marine air support. Employing de Havil-
land D-4 planes, Marine pilots engaged in an angle diving technique to
enhance bombing accuracy. [22]

The Corps' endeavor to tie their destiny to the fleet remained ever elusive,
however. As economic hard times savaged government revenues the Hoover
administration called on the armed services to cut spending. General Douglas
MacArthur, Chief of Staff of the Army and Admiral William Pratt, Chief of
Naval Operations, entered into a quiet understanding. The U.S. Army would
take over the Marines. It was an open secret that the Army Air Corps wanted
to absorb Marine aviation as well. Both proposals were never consummated,
but one could speculate that the Marines must have had the impression that
the Navy regarded the corps as expendable. [23]

That perception was hardly dispelled when Admiral William Pratt reorganized the fleet in 1930. Originally assigned as an adjunct to amphibious
warfare, Pratt designated the Control Force as the Submarine Force. Pratt did
offer the Marines a consolation of sorts. He assigned amphibious operations
to the fleet's Scouting Force, later replaced by the Special Service Squadron.
The latter squadron consisted of a motley collection of aged ships designed
to show the flag in Central and South America. Earlier, when the Bureau of
Navigation (personnel) offered Admiral William Pratt command of the Special Service Squadron, he turned down the assignment on grounds it was
tantamount to a career dead-end. [24]

It was almost as if the Navy did not quite know what to make of joint operations or what to do with Marines and their doctrine of seizing an enemy island by frontal assault. Nor did the Marines find it reassuring that the Naval War College dropped amphibious operations as a course of study. In the meantime, the Navy's Orange plan adhered to its orthodox doctrine. In the event of a Japanese war the Battle Fleet would transit the Pacific (the "through ticket"), engaged and overwhelmed the Japanese fleet in a super-Jutland engagement. Joint operations appeared as a collateral footnote.

LOGISTICS ASHORE

In the 1920's and early 1930's, naval logistics touched on five activities; naval oil reserves, fleet self sufficiency, refueling at sea, merchant ship construction and fleet exercises.

Petroleum Reserves

As the Navy modernized its battleships, fuel oil now emerged as an indispensable commodity. The Navy had secured oil naval reserve #1 and #2 under the Taft administration and reserve #3 under the Wilson administration. [In 1923, Warren Harding designated Alaska's North Slope as oil reserve #4.]

By 1924, Interior, on behalf of the Navy, had selected the Pan American Oil and the Mammoth Oil Company to lease Naval reserve #1 and #3 respectively. The contracts called upon each leasee to exchange crude for refined oil, to construct pipelines and to fill oil tanks at Pearl Harbor and both U.S. coasts.

One fly remained in the petroleum ointment. Albert Fall, Interior Secretary, accepted money from the Pan American Oil Company. In a letter to a congressional committee he denied any financial transaction had taken place. By January 1924, Senator Thomas Walsh, Public Lands Committee, uncovered Fall's blunder. Fully aroused, the U.S. Senate charged fraud and searched for heads to roll. Fall had left the government; Harding had died in August 1923; Denby stood alone.

Denby had testified on the navy's oil lease contracts before the House Appropriation Committee, outlined the Navy's oil storage plan in his 1922 Annual Report, supplemented by reports from the Chief of the Bureau of Engineering and the Bureau of Yards and Docks. In October 1923, the Secretary was called to testify before the Senate Public Lands Committee. Released from New York hospital prior to his appearance, the Secretary's performance was less than inspiring. The Secretary attempted to field Walsh's questions as the Senator walked through a stack of papers, memos and documents. Unprepared, Denby asked to return to the committee after he had time to examine relevant documents and files. Senator Walsh replied that a second visit to the committee was unnecessary. The record, said the Senator, showed that Denby didn't know what he was talking about.[25]

After Fall's misjudgement had come to light, Walsh introduced a Senate resolution calling upon President Coolidge to cancel and rescind the oil leases. As the oil controversy enveloped Congress, a few Senators castigated Denby as a "Siamese" twin of Albert Fall. In January 1924, the Senate took action. Senator Joseph Robinson, Minority Leader, introduced a resolution calling upon President Coolidge to remove Denby from office.[26]

On January 12, 1924, Edwin Denby, his wife, and friends left a Washington theater at around eleven P.M. A White House secret service agent approached the group and instructed Denby to attend a meeting at Senator Lodge's home on Massachusetts Avenue. Denby departed alone. He arrived at the Lodge's home and was ushered into the Senator's library. There he met Senators John Lodge, Irving Lenroot, George Pepper, Summer Curtice, and Frederick Hale. Lodge offered the secretary a cigar. Denby thanked him and Lodge responded that Denby's gratitude was premature. For the sake of the Republican Party Lodge said that Denby should step down as Secretary.[27]

The meeting lasted nearly two hours. In the end, Denby finally put the question to the group. Did they all agree that he should resign? The Senators were ambivalent. They did not ask him to resign but said he should step down. Denby returned to his home after 1:00A.M., and was informed that the White House had called asking where he could be reached.[28]

The Senate debated the merits of the Robinson resolution in late January. One group of senators asserted that Denby was in league with Albert Fall; that he had turned over valuable oil reserves to private oil companies; that he had permitted the Interior Department to let bids without competitive tendering; that he had convinced President Harding to engage in an illegal transfer of Navy oil reserves to the Interior department. Denby's defenders replied that dismissing the Secretary without a hearing before the public lands committee had completed its final report and action violated Denby's right of due process; that the nation's court system, not the Senate, should determine the legality of Harding's presidential order; that the Wilson administration had set the precedent of leasing Naval oil reserves to private oil companies; that Denby had duly briefed Congress on his oil reserve policy; that Denby's actions had adhered to Congress's law as amended by the Daniels' rider.[29]

A few Senate members demanded a bill of impeachment. Senator Robert LaFollette reminded the Senate that that prerogative was reserved to the House, not the Senate, a comment that prompted a few members to approach the House of Representatives. Senator Walsh thought such formal hearing was unnecessary. He put his case succinctly; you can't impeach a cabinet officer for stupidity. Thomas Walsh simply wanted Denby out as Secretary.

On February 11, 1924 the Senate voted on the Robinson resolution calling for Denby to step down.[30]

Meanwhile, two Detroit attorneys arrived in Washington, D.C. to give legal aid to the beleaguered Secretary. They asked to examine naval and Interior leasing documents. Denby turned to War Secretary John Weeks for advice. Weeks told Denby to check with the President. Denby made such an appointment. President Coolidge refused the document request, adding that his lawyers were looking into legality of the Navy oil lease arrangements.[31]

The fight went out of Denby when, in a subsequent White House visit, the Secretary discussed a forthcoming Navy Day celebration. The Secretary asked the President to approve a suggested guest list, which, presumably, included Denby and his wife. The President replied that his wife would handle the guest list. Denby left the White House convinced that President Coolidge did not want him to attend the event. On February 18th, Denby penciled his resignation and hand delivered it to the President. Denby recalled that Calvin Coolidge appeared relieved by the Secretary's action. As they walked to the White House door, the President asked Denby to remain inside and they chatted for the next two hours.

In the interim, President Coolidge picked up the oil reserve cadence. He appointed two government prosecutors who filed a complaint against Harry F. Sinclair of the Mammoth Oil Company and Edward Doheny of Pan American. By this time, Secretary Wilbur Curtis had banned Rear Admiral Robison, Engineering Chief, from making any reference to the Navy's orange war plan. Admiral Robinson's lips, on the stand, were thus sealed. When Doheny's defense counsel attempted to probe into the Navy's Pearl Harbor defense plans, Rear Admiral Robison replied he was under orders not to answer.[32]

The Senate was not yet finished with Admiral Robison. When his name appeared on the Navy's list for the permanent rank of Rear Admiral, Senator Thomas Walsh insisted that Robison's name be stricken. The Navy obliged. President Coolidge did not promote Robinson and he retired with the rank of Captain. There was now a talk of Thomas Walsh running for President.[33]

Denby set his resignation date effective March 10, 1924. In the next few weeks Denby's memos to the file revealed actions that bordered on the surreal. Senator Thomas Walsh asked if he could employ the Navy's Intelligence Service to investigate the Navy Department. Denby replied that department policy precluded such use. (Intelligence resided within the office of the Chief of Naval Operations.)[34]

The oil investigations wound their way through the nation's court system, reaching the Supreme Court in 1927. The court ruled that Albert Fall had initiated the leasing contracts with Pan American and the Mammoth Oil companies, absolving Edwin Denby of any wrongdoing. On the other hand, the court rejected the Navy's Judge Advocate General interpretation that

Congress had empowered the Navy Secretary to buy, sell, exchange or store any of its reserve oil. That authority, stated the court, exceeded the will of Congress. The court held that the Navy could retain $500,000 worth of oil receipts, the remaining revenues had to be turned over to the Treasury Department. The court thus sustained legislation that Congress did not write, failed to read, blindly cast a vote, denied they cast a vote, and subsequently disclaimed its meaning and intent. [35]

More important, the actions of Congress, the Navy, and Interior, rested on the supposition that the U.S. was in the midst of a national oil famine. As early as 1916, the Interior Department had predicted that the United States would exhaust its oil reserves by the late 20's. Oil conservation was now considered a matter of national urgency. [36]

In 1924, prospectors discovered oil in east Texas. The strike proved a mother load. Oil flooded the market. The supply was apparently unstorable and unstoppable. By the early 1930's the price of oil had plunged from $3.50 to 10 cents a barrel. An oil shortage had thus confounded Presidents Taft, Wilson, Harding, Coolidge, and Hoover. The dire predictions of the Interior Department had misled Secretaries Meyers, Daniels, Denby, and Wilbur. The actions of government petroleum experts and Congressional staff members were refuted by unlettered wildcatters seeking to turn a dollar in Eastern Texas and Oklahoma. [37]

President Coolidge proceeded to create a Federal Oil Conservation Board (FOCB) in 1924, a board that included the Secretaries of the Navy, Commerce and Interior. The Board took as its mandate to cut U.S. oil output. As a move to stabilize oil prices individual states were encouraged to adopt production quotas. [38]

Two juries, in the meantime, held that Edward Doheny and Harry Sinclair were innocent of offering a bribe to Albert Fall. The same court, the same judge, a different jury, sent Fall to prison for accepting a bribe. Unable to meet his loan payments to Doheny, Albert Fall forfeited his New Mexico ranch to Mrs. Estelle Doheny. In 1927, the Supreme Court invalidated the Navy's oil leasing contracts. In the meantime Interior had predicted that Teapot would yield 130 million gallons of crude oil. After producing 3.5 million barrels, Teapot Dome ran dry and was capped. Teapot Dome turned out to be a dry hole. But all was not lost, Teapot Dome generated grazing fees from western cattle ranchers. [39]

Logistics Afloat

Admiral Robert E. Coontz and Admiral Edward W. Eberle exchanged commands in mid 1923. Eberle became Chief of Naval Operations; Coontz, Commander-in-Chief of the U.S. Fleet. In December of that year, Admiral Eberle, outlined of the concept of a mobile base that included the fleet train and the Naval Transportation Service. In the event of an outbreak of hostilities, a mobile supply unit would accompany the battle, control, base and submarine force across the central Pacific. The Navy contemplated setting up floating bases at anchorages at Eniwetok, (Marshalls), and Ulithi (Carolines). From those anchorages, the fleet would proceed to Guam and/or the Philippines. The mobile base plan also contemplated four classes of floating drydocks capable of servicing the displacement of battleships, cruisers and destroyers. Admiral Eberle put the matter directly, "a dry-dock may save a ship and thus may have a deciding influence in a campaign."[40]

As Commander-in-Chief of the U.S. Fleet, Admiral Robert Coontz insisted that the fleet should move toward a goal of fleet self-maintenance. Coontz noted:

1. the development of the Fleet Base Force upon the axiomatic principle that it is one of the four major subdivisions of the feet.
2. the indoctrination of the fleet with the realization of the practical difficulties encountered by the shore establishment to the end of that maximum cooperative effort of the shore establishment in affecting and maintaining fleet readiness may be attained.
3. development of the train to the end that it may refuel, re-victual, restock and repair combatant units on the high seas.[41]

In summarizing fleet logistics, Admiral Coontz wrote further:

Fleet re-fueling and re-victualing doctrines have been developed and are now in operations. They encompass operating conditions similarily for peace or war; and increase the Fleet readiness and radius through decreasing the turnaround of Train vessels.

Fleet self-maintenance has been increased through study of periodic issues of consumable supplies; and a system has been established whereby that Force in the Fleet organization which is charged with the establishment of an outlying temporary war operating Base in the field of active operations, is enabled to envisage its requirements. It yet remains properly to stock this Force with the supplies. This development should be vigorously proceeded with until the point is reached where the Trains of the Fleet Base Force can and do completely and continuously supply all requirements of the active Fleet. In no other way can the reserve of material necessary to stock a temporary Base be efficiently accumulated afloat, and until it is actually afloat the Fleet will be unable to move suddenly with confidence. Its consummation will eliminate the desire to establish permanent Bases on shore in localities frequented by the Fleet, and remove that source of irritation due to not being able to obtain supplies from present short sources in the limited times desired.[42]

While the Navy was fleshing out its mobile base project, Secretary Denby emphasized the fleet's reliance on the U.S. merchant marine industry. In the event of a national emergency, Denby observed that the fleet,

> ...must draw from the merchant marine for Navy use – cruisers, transports, and fuel ships, repair ships, distilling ships, hospital ships, mine layers, mine sweepers, patrol ships, destroyer tenders, submarine tenders, aircraft tenders, tugs and supply ships of all kinds.[43]

Denby was essentially running down the components of a mobile supply force.

Refueling at Sea

U.S. fleet exercises continued through the end of the Hoover Administration. War plans, Fleet Training, the Bureaus, the Base Force, all would play a role in developing refueling exercises. In the winter exercises of 1923-1924, for example, the oiler *Kanawha* employed a broadside technique and refueled two destroyers at the same time—a protocol not entirely without risk. Whether by sudden wave action, disabled rudders, station miscalculation or simply poor communications, ships could and did collide. Damaged vessels could easily jeopardize an officer's chances of promotion.[44]

The astern method, on the other hand, posed as an alternative to the broadside technique. Here vessels travelled in column, the oiler pulling the destroyer, the process reversed in the case of a heavy warship. The astern protocol did reduce the incidence of ship damage, but the rate of oil transfer was slow and fleet oilers could refuel but one ship at a time.

In 1925, Admiral Robert Coontz embarked upon an ambitious project. Concerned that traditional fleet exercises were overly short in duration and hence no real test of fleet self-maintenance, Coontz commanded the fleet from the U.S. to Australia. In recalling the experience to the U.S. War College, Coontz rendered an optimistic assessment of the navy's cross-Pacific experience. The fleet, said Admiral Coontz, had mastered replenishment underway.[45]

The Australian cruise also served as a platform to test naval wireless equipment. Dr. A. Hoyt Taylor, NRL scientists, traveled aboard the *U.S.S Seattle*. Fleet radio operated on long continuous wave spark transmitters. Fred H. Schnell, a member of the American Radio Relay League (ARRL), and a short wave expert, travelled aboard the Seattle. The Navy asked Schnell to take a leave from ARRL, assume the rank of a Reserve Navy Lieutenant, and operate recently installed short wave equipment. During the seven month cruise the long wave sets acted erratically, the short wave sets proved surprisingly reliable. Following the voyage, the Navy embarked on a program of radio upgrading.

In 1927 the oiler *Kanawha* fueled the battleship *Arizona* employing the astern method. The oil flow rate approached 39,000 gallons an hour. As part of the exercise, the *Kanawha* heated fuel oil to 140 degrees Fahrenheit, increasing the rate of fuel transmission. Intrigued by the possibility of reducing refueling time, War Plans urged even faster speed. Nor were the Navy's refueling exercises confined to oilers and warships. Battleship refueled destroyers; tenders refueling submarines; carriers refueled cruisers. Over time Naval officers began to detect a nexus between fleet mobility and fleet logistics. As one officer put it: "One of the greatest strategic advantages to be possessed is mobility. Very few naval officers realize the potential of a mobile strike force except those that participate in refueling at sea."[46]

In 1930 Operations began to favor broadside refueling, War plans directed the fleet to turn in its associated astern gear and equipment, noting that ships forces could either remove the equipment themselves or they could rely on Navy Yards during a scheduled overhaul.[47]

Naval officers remained dissatisfied with auxiliary speed. Oilers simply could not keep up with fleet. Speed under 10 knots exposed the Battle Fleet to submarine attack. Some officers even insisted that 10 knot fleet exercises were not a proxy for war preparation. Some argued that oiler speed should be at least 12 knots.[48]

The Coolidge Administration did take steps to rehabilitate the nation's merchant marine industry. In 1928, Congress passed a subsidy bill not fundamentally at variance from Harding's failed proposal of 1922-1923. Under the bill, the U.S. Post Office provided a mail subsidy for the operation of fast liners. Congress supported loan insurance program for private ship building yards. The legislation led to the construction of 31 new passenger liners in addition to an upgrade of 41 merchant vessels. Powered by steam turbines, merchant ships lent themselves to naval auxiliaries, if and when needed.[49]

SOURCING

In procurement matters, the Navy's in-house yards, stations, arsenals and factories, found both buyer and seller residing under common affiliation, a vertical tie that permitted negotiations to be conducted by colleagues within the same corporate family. In-house procurement had obvious advantages. It was free of the adversarial give and take associated with dealing with outside vendors. Naval officer procurement was untainted by incentives that energized the world of commerce. The former was impelled by public service; the latter animated by corporate profits. (An economist would be awarded a Nobel for observing that government and private personnel were equally activated by self-interest.)

In 1929, Congress legislated that over half of all cruisers had to be constructed in government yards. President Hoover also assigned destroyer construction to three eastern government yards in order, "to give steady employment to our faithful employees." By this time the Navy perceived civilian yard employees as members of its shore establishment family.[50]

By the early 30's Congress had stipulated that ship repair constituted a Navy Yard monopoly; allocated 50% of cruiser construction to naval yards; banned ship forces from doing repair work while docked at a navy yard; discouraged the introduction of modern production methods; prohibited time and motion studies; created a torpedo supply monopoly; reimbursed a navy yard for cost over-runs; imposed no penalty for tardy ship delivery; promoted yard diversification into non navy work; permitted Navy Yards to submit estimated costs in their construction bids; blanketed yard employees with civil service tenure; and accorded employee wage and benefits packages that exceeded the compensation of private shipyards. Under financial duress, Newport Shipping and Dry Dock, Inc. diversified into locomotive repair work. Washington politicians insisted that government yards existed to keep private yards honest. Put differently, Congress, the Bureau Chiefs, and civilian yard unions sat on the same side of the table.

The nation's plane manufacturers experienced their own financial disincentives. Once a firm submitted a prototype design for naval consideration, the plans automatically became government property. The experience of the Curtiss Aviation Company spoke volumes. The Bureau of Aeronautics to Curtiss's prototype and solicited competitive bids. The Curtiss Aviation Company bid $32,000 on its own design aircraft, Martin, a rival, submitted a bid of $20,000. Curtiss not only lost the contract bid but found that its research effort had funded a market competitor.[51]

Congress curtailed the buying authority of the Secretary of the Navy by legislating the Aircraft Procurement Act of 1926. A Navy Secretary may have been attempted to engage in negotiated contracts, but the Navy's Judge Advocates Office, backed by Comptroller General, held that the Secretary had no such discretionary authority. A Secretary was mandated to accept to take the lowest, sealed bid. The consequences may have been unintended, but congress's restricted procurement stifled and curtailed naval aviation research and prototype development.[52]

To be sure, congress had adopted a five year aviation appropriations plan after 1926. But private aviation firms continued to experience regulatory oversight. Government inspectors roamed plants to determine whether manufacturing expenses were reasonable and properly allocated. Assembly plants thus became subject to audits not unlike those experienced by a regulated public utility. Moreover, Congress insisted that a 20 year amortization rate was appropriate for corporate tax purposes. By the mid 1920's, the combined employment of the army's McCook Field, the Navy's aircraft factory and the

Pensacola Air Station, exceeded the employment of all aviation firms in the nation's private sector.

In the late 1920's, a problem thought to have been resolved erupted again. Navy built submarine diesels continued to breakdown, imposing a limit on boat speed and operating range. The problem was somewhat surprising. Navy diesels had been essentially carbon copies of German U boat diesels employed during the First World War. Admiral H.S. Yarnell, after a visit to Germany in 1930, reported that German firms had adopted new metallurgy into its propulsion units, achieving a 30 pound per horse power ratio. [53]

Back in the States, Admiral Yarnell explored the buying choices open to the department. The department could license M.A.N. diesels for U.S. production, an option he said, would hardly invite a warm reception from Capitol Hill. As a second choice, the Navy could buy German M.A.N. engines outright and install them in U.S. boats. Admiral Yarnell predicted that such a "buy" choice would ignite "....an outcry from yard labor unions." Yarnell added if that the Navy "...tried to build the engines outside we would have the whole congressional delegation from New York on our backs." [54]

Stymied and frustrated, Admiral Samuel Robinson approached Charles F. Kettering, a General Motors vice president. Robison asked if the company could develop a light weight, diesel engine capable of an operating radius of 12,000 miles. Kettering came up with a twelve cylinder "pancake" diesel. The diesel eventually reached 27.5 pounds per horse power. Although the power plant experienced the usual teething problems, GM's diesel proved durable and reliable. [55]

In a similar vein, the Marine Corps began searching beyond the Bureau of Construction and Repair for an answer to its landing craft needs. As early as 1928, Andrew Jackson Higgins of New Orleans touted a flat boat used for rescue work in the Louisiana bayous. Marine's interest was dampened, however, by tight budgets.

Private Sector Incentives: the Buy Option

Less than one year in office, the Hoover Administration was blindsided by New York's stock market crash. As the nation's output dropped in half, prices fell, consumer spending plummeted and corporate earnings evaporated. Over 20% of the nation's labor force would experience involuntary unemployment. The length, depth and cause of the Great Depression remains controversial to this day. Several candidates, nevertheless, were said to be causative factors; the stock market, the Federal Reserve System, the U.S. banking industry, U.S. fiscal policy, trade protectionism.

There was, of course, nothing unusual about the nation's business cycle. The nation had endured financial fluctuations in the past. However severe, all were regarded as short term and self-correcting. A few business cycles scholars insisted that the 1929 crash was exacerbated by speculation associated with margin stock purchases. Things went swimmingly well when the market went up; but a downturn proved self-reinforcing. Falling stock resulted in margin calls that precipitated of flood of selling which fed another round of margin calls. Economists argued, however, that stock fluctuations did not cause the depression.

A second school of thought pointed to the policies of the Federal Reserve. The U.S. Federal Reserve reduced credit by 30% and raised U.S. interest rates to compete with the U.K. (France, in the meantime, hoarded gold without expanding its money supply.) Viewed in nominal terms U.S interest rates of five percent appeared relatively modest. But as prices fell the real rate of interest soared as high as 16%. High real interest costs curtailed investment expenditures and business capital spending began to fall off, all contributing to slow growth and involuntary unemployment. [56]

Critics also added that the Federal Reserve failed to support the nation's banking industry. Originally, the reserve system had been created as a banker's bank. As the panic spread, depositors lined up to withdraw their funds. Unable to call in their loans and running out of cash, banks closed their doors.

Government fiscal policy was regarded as still another depressing factor. As industry laid off workers, incomes plummeted, tax receipts fell, prompting the federal government to cut expenditures and raise taxes. The combination of tight monetary and higher taxes, became a lethal mix. Presumably federal deficit spending would ripple through the economy by boosting income, output and employment. Federal deficit spending, however, was countered by state government spending cuts, who by law, were required to balance their budgets. Fiscal policy was not without its limits.

To many economists, U.S. tariff policy stood as prime source. As U.S. business sales fell, Congress raised tariffs on foreign competitors. Overseas firms, unable to generate income from U.S. sales, were in no position to buy U.S. exports. Foreign governments retaliated by boosting their own tariffs. Protectionism begat protection as world trade began to taper off.

On the other hand, the nation had experienced higher tariffs in the past and survived the experience. And as a percent of U.S. total output imports/ exports were relatively small. A general consensus emerged that import tariffs did not "cause" the depression, although, protectionism did generate economic retaliation. [57]

President Hoover was convinced that wage competition had depressed consumer spending. The government's buying power, he thought, could address the problem. In 1931, Congress legislated the Davis-Bacon Act mandating that government pay the higher prevailing wage rate to workers in a given geographic area. The law attempted to prevent labor competition in bidding on schools, post offices and court house construction. The prevailing wage, however, turned out to be a tax imposed upon an already over stretched ratepayer. [58]

A few members of Congress asserted that market failure was the real key. The market, they asserted, was neither self-adjusting nor self-correcting. Total demand was insufficient. The economy was stuck. The government had to step in, encouraging the unemployed to dig holes and, if necessary, refill them. Keynes and fiat money now stalked the world.

A positive side did emerge from the 1920's, however. The nation's industrial base expanded as illustrated by the electric power industry. From a base of 17 million miles of electric power line more than doubled in the decade of the twenties. Power station plant and distribution facilities grew from $800 million dollars to $2.8 billion dollars. [59]

Electricity transformed U.S. manufacturing practices as small motors replaced belt driven reciprocating machines. Corporate research led to the development of high pressure steam boiler and turbine technology. Power stations cut unit costs as utilities adopted electric meters to segment users through off-peak pricing. Investment, innovation, scale economies combined to reduce electric rates to the consuming public.

Exhibiting a voracious appetite for capital, electric utilities extended service over the broad reach of the Midwest. State commissions exercised oversight over plant investment, earnings, expenses and rate payer tariffs. Not unlike Midwest grain elevators, U.S. courts defined power suppliers as a regulated utility.

By the 1920's corporate research laboratories emerged as a competitive tool. Westinghouse, General Electric, Allis-Chalmers, stationary turbine suppliers, generated high steam pressure of 600 lbs. per square inch. Westinghouse approached the Navy on the turbine sales. The department responded that its maritime suppliers met its needs satisfactorily. Something new was afoot, nevertheless. A product developed in a commercial market posed as a potential application in the military market. [60]

Although the Coolidge and Hoover Administrations attempted to discourage government agencies from competing with private sector firms, the legacy of the Wilson Administration remained ever inviting. During the First World War Congress approved construction of a hydroelectricity plant for the purpose of nitrogen production. Several lawmakers wanted the federal government to enter the electric power market, but Presidents Harding, Coolidge, and Hoover resisted that option. [61]

The sales of consumer durables, refrigerators, vacuum cleaners, air conditioners, electric irons, fans and radios, exploded in the 1920's. In 1921 consumers spent $65.2 million on electrical appliances; by 1929 that expenditure level had trebled, activated by installment buying. Buoyant demand enlarged the nation's commercial infrastructure. [62]

Coolidge and Hoover adopted policies hospitable to the development of the radio industry. Commerce Secretary Herbert Hoover allocated frequencies as a device to alleviate radio interference. RCA had applied for an exclusive wireless telegraph license to China. Although Denby saw merit in the company's application, he was adamant that the domestic radio industry remain open to competitive forces. In the meantime, amateurs and ham radio operators developed the short wave radio market that in turn nourished the growth of radio component and accessory suppliers. By 1930, the cathode ray tube signaled the beginning of television. [63]

The Navy, as noted above, had anticipated that the GE-RCA radio structure would to impose order on a competitive wireless market. For a short time, the arrangement appeared workable. When amateurs moved into radio broadcasting, the spread of networks converted GE, RCA, Westinghouse and AT&T into economic rivals. Under competitive threat, clean distinctions between radio, telephone and telegraph markets began to soften and erode. In 1926, AT&T announced that it would sell its broadcasting network to RCA. By 1930, some 60 radio broadcast networks operated in the U.S. domestic market. [64]

The nation's telephone industry's wire line investment doubled from 15.3 million miles to 36 million miles during the 1920's. Not unlike, the electric power industry, the telephone industry segmented customers by geographic and service categories. A telephone holding company of some 23 operating affiliates, AT&T's long distance network provided toll links to the company's local affiliates. Independent telephone companies set up local service in the less populated sections of the country.

Demand for residential and business voice service, in turn, spilled into the telephone equipment and related apparatus market. Phone rates declined, usage expanded, stimulating improvements in terminal stations, manual and automatic switching equipment, and transmission lines. AT&T's manufacturing affiliate, Western Electric, provided cable, telephone sets and switching gear to its Bell operating companies. Some dozen equipment firms delivered apparatus to the independent voice market. In 1931, AT&T acquired the Teletype Corporation as a Western Electric subsidiary. Radio, telegraph and telephone technology began to overlap.

The twenties heralded the growth of automobiles. The Ford Motor Company dominated the industry while General Motors and its random assembly of auto affiliates came perilously close to insolvency. After DuPont acquired controlling interest, General Motors placed Alfred Sloan as the company's chief executive's officer. Sloan segmented automobile lines along customer income, eliminated duplicate operations, and via a series of committees coordinated the financial operations of the company. By 1930, General Motor Company had surpassed the Ford Motor Company in sales, earnings and market share. Sloan's organizational gift, in fact, compelled Ford to abandon his Model T production, to shut down the plant for nearly a year and invest $18 million dollars installing 4,500 machine tools in preparation for its model A.[65]

Automobile sales stimulated the demand for steel, copper, lead, chemicals, lacquer, and rubber, glass production, to say nothing of broaches, grinders, and stamping, drilling, and boring machines. Gas stations, restaurants, and hotels sprung up along motor ways as individual states picked up the burden of road construction, 370,000 miles of paved road in 1920 doubled by 1929, accompanied by the sales of graders, trucks, concrete, and asphalt.

Petroleum consumption tracked automobile registration. Pre-ignition engine firing inspired GM and Standard Oil of New Jersey to form a joint venture to sell a gasoline additive (tetraethyl lead). The oil industry expanded its pipeline network and placed refining plants along the mid-continent and the U.S. northeastern corridor. The industry's tanker fleets and railroad tanker cars further enlarged the nation's petroleum infrastructure.

Congress permitted Walter Brown, Hoover's Post Master General, to create a national airline system in 1930. Brown designated American, United, TWA, and Eastern as continental carriers. The Post Office tied a mail subsidy program to plane capacity as apposed to mail weight. Brown insisted that planes be powered by multiple engines and pilots subject to examination and licensing.[66]

The innovation of air cooled engines paralleled the development of stressed fuselages, engines cowls, wing flaps, retractable landing gears, two way radios, and adjustable pitch propellers. In 1929, U.S. passenger miles flown were just short of 170,000 miles. By 1933, the number had exceeded 500,000 miles. Hoover's program of safety, pilot licensing, and flight navigation aids was beginning to yield results.[67]

Throughout the 1920's, the aviation industry sponsored national races. By the end of the decade private firms had introduced monoplanes capable of speeds in excess of 230 miles per hour, propelled by 400-horse power air cooled engines. Private planes now exceeded the performance of military aircraft. Unable to compete military pilots dropped out of contention.[68]

Aviation firms adopted the holding company as a business model by the late 20's. The United Transport and Service Company, for example, operated a coast to coast passenger service, integrating Boeing's manufacturing operation, and Pratt and Whitney as supply affiliates. The AVCO Corporation emulated the united corporate structure. Henry Ford, spurred by "Lindberg Fever," entered the commercial aviation field in 1927.[69]

Aviation executives sold stock to the general public in the late 1920's, liberating firms from commercial bank loans. A few plane manufacturers accepted equity in lieu of cash payment from financially stretched airline carriers. Other firms husbanded their earnings as a hedge against the nation's business cycle.

Recovering from World War I, the railroad industry now found themselves competing with cars, trucks, buses and oil pipelines, alternatives that diluted rail revenues. The nation's farmers were dependent upon world grain prices. In general, the era of the twenties not only experienced an expansion in consumer and nonconsumer spending but deepened the nation's industrial infrastructure.

The economy experienced a second industrial revolution. Taylor's scientific management doctrine spread throughout the nation's industrial sector. Ford and General Motors employed time studies, inventory control, standardized machinery, the assemble line. Nor was the auto industry the only industry receptive to a new management doctrine. General Electric, Westinghouse, Western Electric, Bethlehem International Harvester employed facets of Taylorism that lifted productivity, reduced unit costs, funded wage increases, raised profits and passed lower prices forward to the consuming public. Scientific management was accompanied by the growth of consulting firms - Arthur D. Little, Booz Allen, Andersen, Gilbreth, Gantt, McKenzie et al. More important Taylor's writing were translated and his practices adopted in Europe and Japan.

The U.S. was singularly prosperous during the decade of the twenties. The application of science to U.S. management, the evolution of specialized machinery, a new class of administrative staff, the emergence of the assembly line, the standardization of piece parts, refined cost accounting, efficiencies gained on the shop floor, mass distribution, and price discounts converted luxury products into consumer necessities. John Braeman summed up the roaring twenties by observing that "...the United States produced an output of manufactures larger than that of the other six major powers – Great Britain, Germany, France, The Soviet Union, Italy and Japan combined."

The 1920's was indeed a heady decade. [70]

INDUSTRIAL MOBILIZATION POLICY (IMP)

By the end of the twenties, the Army Navy Munitions Board (ANMB) had completed a U.S. mobilization plan in the event of a national emergency. The 1930 plan contemplated a central agency responsible for coordinating U.S. industrial production. The agency, a War Resources Board, and its administrator would balance the needs of the military and the civilian economy. The ANMB recommended the civilian administrator be conversant and experienced in industrial management. The ANMB also suggested that the War Resource Board be given authority to regulate industrial prices and corporate profits.

Congress, in response to the plan, set up a War Policies Commission. The commission invited testimony from interested parties, including General

Douglas MacArthur, the army's chief of staff, a Navy representative Captain H.K. Cage, and Bernard Baruch, former head of the War Industries Board.

General Douglas MacArthur testified that a military officer be appointed as director of a War Industrial Board, empowered to regulate the economy's prices and corporate earnings. The Army's Chief of Staff, saw no need to require labor conscription. [71]

Captain H.K. Cage, U.S. Navy, backed the Army's overall plan. One lawmaker asked Captain Cage if the Navy Department had assessed the details of MacArthur's plan. Captain Cage responded "...the Navy department has had no chance to carefully study the statement and must therefore make no comment." Another Senator questioned whether the Navy had developed its own Industrial Mobilization Plan, whether the General Board had seen the Army's ANMB plan; and whether the navy's General Board had approved the ANMP's plan? Captain Cage responded no to all inquiries. [72]

The Commission heard next from Bernard Baruch. Baruch agreed that, ANMB's proposal to create an overarching agency as mobilization coordinator. Baruch had reservations concerning over the board's membership. Cabinet officers should be excluded. They had enough to do simply running their own departments. Baruch said it said was imperative that a civilian be selected who had demonstrated competence and expertise in the field of industrial production. A military officer, by definition, was ill-qualified for such a post. [73]

Completing its hearings the War Policies Commission submitted its recommendation to congress in 1931. The Commission concluded that any authority to regulate private sector prices and profits required an amendment to the U.S. Constitution. The Commission opted in favor of a board composed of U.S. cabinet officers. The commission agreed that labor conscription need not be required. In a word, Congress essentially ignored the Commission's findings.

INTER-GOVERNMENT COORDINATION

As part of general reorganization, President Harding proposed the formation of a Department of Defense (DOD). After Harding's death, Congress held hearings on the plan. Secretaries Weeks and Denby testified against the DOD concept, apprehensive that a merger would create an independent aviation unit. Both cabinet officers sought to retain their air unit. The two Secretaries need not have worried. Calvin Coolidge viewed executive reorganization as a matter of low priority.

Herbert Hoover, on the other hand, remained committed to reorganizing the government's Executive Branch. For one thing, Hoover attempted to divest nonmilitary activities from the Navy and War Department. But Congress had other plans. As Dr. Lawrence J. Legene put it; a "lame-duck Democratic House on January 1933 passed a resolution 'vetoing' the entire lot."[74]

The question of cooperative military command remained unresolved in the 1920's. By late 1932, the U.S. Army Air Corps favored aviation consolidation, a department of defense and a government agency responsible for civil and military aviation. Naval officers apposed a defense department concept, insisting that the President, as Commander in Chief, had sufficient power to administer the nation's Armed Services. There the matter would rest until the events of 1941.

NOTES

1. Thomas P. McGruder, "The Navy and the economy," *Saturday Evening Post*, 99, No. 9 (September 24, 1927): 148.

2. NARA, RG80, General Board, Hearings, 18 January, 1932, "Naval Policy Regarding Research and the Naval Research Laboratory," 109-112; also Ivan Amoto, *Pushing the Horizon: Seventy Five Years of High Stakes Science and Technology and the Naval Research Laboratory* (Washington: 1998), 72.

3. John B. McKinney, Radar: A Case History, *IEEE A and E Systems Magazine*, 21, 8 (August, 2006), 41; also David Kite Allison, *New Eye for the Navy: The Origen of Radar at the Naval Research Laboratory* (Washington: Naval Research Laboratory 1981), 6.
U.S. Congress, Aircraft, House Committee on Interstate and Foreign Commerce, Hearings on the President's Aircraft Board, September, 1925.

4. U.S. Congress, Congressional Record, 68th Cong. 1st. Sess. House of Representatives, March 15, 1924, 4262, Vol. 45, Roll 164.

5. U.S. Congress, Congressional Record, 68th Cong. 1st. Sess. House of Representatives, March 20, 1924, 4589, Vol. 45, Roll 164; also Captain A.W. Johnson, *A Brief History of the Organization of the Navy*, Department, March 1933, p. 135.

6. U.S. Congress, Navy Department Appropriation Bill, 1925, Hearings Before Subcommittee of House Committee on Appropriations, 68th Cong., 1st. Sess., December 10, 1923, 69. Also Stephen C. Suonavec, "Congress and the Navy: The Development of Naval Policy, 1913-1947," Ph.D. dissertation, Texas A and M, 2000, 145.

7. Calvin Coolidge, Coolidge papers, Roll #21, Letter to Secretary of the Navy, Lamont Library, Harvard University, February, 21, 1924.

8. NARA, RG 80, U.S. Navy, General Board, Special Board, January 17, 1925, 75 Micro #280, Reel 6. "The Battleship is the element of ultimate force in the fleet, and all other elements are contributory to the fulfillment of its function as the final arbiter in sea warfare."

9. Rear Admiral W.F. Fullam, "The Passing of Sea Power," *McClures*, 56 (June 1923): 30.

10. Paolo E. Coletta, *The American Naval Heritage* (New York, University Press of America, 1987), 273.

11. NARA, RG 80, Hearings of the General Board of the Navy, 8 July 1930, Man Engines and Necessary Auxiliaries for the U.S.S. V-8 and U.S.S. V-9, 253.

12. NARA RG 313, Annual Report, Commander in Chief, Battle Fleet, 1 July, 1926, 64-65.

13. Eugene E. Wilson, "The Navy's First Carrier Task Force," U.S. Naval Institute Proceedings, 76, 2, (February, 1950): 164.

14. Vincent Davis, *The Politics of Innovation: Patterns in Navy Cases* (Denver: University of Denver, 1967), 47-48, Columbia University, Oral History, Eugene Wilson, 444 "...I knew that I had jeopardized my career. The old timers that resented aviation would think me a renegade...."

15. Forrest Sherman, "Some Aspects of Carrier and Cruiser Design," United States Naval Institute Proceedings, 80, 11, (November, 1930): 997.

16. Samuel M. Robinson, Oral History, Columbia University, "The development of a light high speed diesel engine was dear to my heart," also H.L. Hamilton, "Historical Record and Notes of the Development of Electro-Motive," Kettering University, GMI Alumni Foundation Collection, Flint, Michigan, no date.

17. NARA RG 80 Office the Chief of Naval Operations, Fleet Training Division, General Correspondence, 1914-1941, Box. No. 1, memorandum from Bureau of Ordnance on Force Torpedo Bulletin, No. 1, 16 October, 1923.

18. NARA, RG 127, Report on fleet problem No.4, February, 1924, Marine Corps University archives, report of commander-in-chief, U.S Fleet, radio and visual signal equipment, item 48; also, Arthur Hezlet, *Electronics and Sea Power* (New York: Stein and Day, 1975), 161-162; see Timothy Wolters, "Managing a Sea of Information Shipboard Command and Control in the United States Navy, 1899-1945," Ph.D. dissertation, Massachusetts Institute of Technology, 2003, p. 176-177.

19. NARA, RG 127, Fleet problem No. 4, February 1924 (Box 14), Material effectiveness report by Commander-in-Chief, "The construction of new carriers should be expedited."

20. NARA, RG 127, Fleet problem No. 4, February 1924 (Box 14), Material effectiveness report by Commander-in-Chief, "The construction of new carriers should be expedited."

21. John A. Lejeune, "Amphibious Landing," lecture, Naval War College, June, 1924, Naval Historical Center, Naval War College, Newport, R.I.; also NARA, RG 127, Records of the U.S. Marine Corps, Division of Plans and Policy, War Plans Section, War Plans, 1915-1946, Box 2. "...the weaker marine expeditionary force is the more support must the fleet give it which means that additional battleships to give support would have to be detached from the fleet which ships thus become unavailable for the battle fleet in crushing the enemy's battle fleet..."

22. E.H. Brainard, "Marine Corps Aviation," *Marine Corps Gazette*, 13, 1 (March, 1928): 30.

23. John R.M. Wilson, *Herbert Hoover and the Armed Forces* (New York: Garland, 1993), 76.

24. Gerald Wheeler, *Admiral William Veazie Pratt: A Sailor's Life*, (Washington: Department of the Navy, 1974), 287-288.

25. U.S. Congress, U.S. Senate, Committee on Public Lands and Surveys, "Leases Upon Naval Oil Reserves," 68 Cong., 1st Sess., October 25, 1923, 1285.

26. *New York Times*, February 2, 1924, 2.

27. Edwin Denby papers, burton Historical Collection, Detroit Public Library, Box 3, dictated memorandum, February 4, 1924 cited as Burton Collection, "When I returned home I found that the White House had called my house and insisted upon knowing where I could be found as the matter was very important." , p. 6.

28. P. 5.

29. U.S. Congress, House of Representatives, Information requested of the Secretary of the navy under H. Res. 204, Edwin Denby to Thomas S. Butler, Naval Affairs Committees, March 7, 1924, 600.

30. Burt Noggle, *Teapot Dome: Oil and Politics in the 1920s*, (New York: Norton, 1965): 150.

31. Burton Historical Collection, Memorandum dictated to John B. May, March 7, 1924.

32. *New York Times*, December 5, 1926, 15, from Secretary Curtis Wilbur to Captain John K. Robison, "...you are advised that confidential information of the government may not be disclosed..." Burton Collection, Edwin Denby to Mrs. J.K. Robison regarding Robison's dropped promotion., 7 august 1923, Box #3.

33. Calvin Coolidge papers, Lamont Library, Harvard University, Navy Board on Selection, 14 June, 1924, Roll #21; finally, *New York Times*, December 5, 1926, 26.

34. Denby papers, University of Michigan, Bentley Historical Library, Memorandum dictated by Edwin Denby to John R. May, 3 March, 1924, 1 [cited as Bentley papers]. Denby states he can employ naval intelligence only on orders from the president, see also confidential memorandum for the secretary of the Navy, 28, February, 1924.

35. 275 U.S. 13 (1927), *Mammoth Oil Company V. United States*, 275, U.S. Supreme Court.

36. *The Oil and Gas Journal*, 40, 53 (January 28, 1921): 2, Secretary Daniels attempted to permit the Navy to commandeer oil at a discount after the 1918 armistice.

37. Gerald D. Nash, *United States Oil Policy, 1890-1964* (Pittsburgh: University of Pittsburgh Press, 1968), 247.

38. p. 117, In Texas, 1927, "The governor then sent approximately 4,000 troops [state militia] into the oil fields to police the closing of wells and to arrest violators."

39. Welton Winas Harris II, "Harry F. Sinclair and the Teapot Dome Scandal: Appearances and Realities, Master's Thesis, Indiana University, 1961, 131. "The total production of the Teapot field, when it was closed down December 31, 1927, was 3, 549, 228 barrels of oil, which was far cry from one of the conservative estimates of 200,000,000 barrels," also P.C. Spencer, *Oil and Independence: The Story of the Sinclair Oil Corporation* (New York: Newcomen Society of North America, 1957), 19.

40. NARA, RG 80, Hearings of the General Board of the Navy, GB420-2, memorandum from senior member to the Secretary of the Navy, April 7, 1923, 10, Box 61.

41. Annual Report, Commander in Chief, U.S. Fleet, 1 July 1923-30 June, 1924, Admiral Robert E. Coontz, 4.

42. 42.

43. Edwin Denby, "The Navy and Merchant Marine are Interdependent, *Marine Engineering and Shipping Age*, 28, 11, (November 1923): 657.

44. NARA, RG 19, Bureau of Construction and Repair, General Correspondence, 1925-1940, memo to Director of Fleet training from H.L. Shenier, February 17, 1928, "Having witnessed the operation of fueling destroyers at sea, I am aware of the attending difficulty and concurrent possibility of accidents in the operation due to the necessity of the exercise of careful seamanship."

45. NARA, RG 19, Rear Admiral Robert E. Coontz, "Logistics," lecture, U.S. Naval War College, Naval Historical Collection, October 7, 1926, 6.

46. NARA, RG 38 Office of the Chief of Naval Operations, Fleet Training Division, General Correspondence, 1914-1941. February 16, 1927, on refueling at sea.

47. NARA, RG 19, Bureau of Construction and Repair, General Correspondence, 1925-1940, from Chief of Naval Operations to Secretary of the Navy, Fueling at Sea, April 13, 1931. "Recent recommendations have favored use of the broadside method even should fueling of a battleship of other large vessel be required..." "It is recommended that all further work on astern oiling gear be stopped..."

48. NARA, RG 38, Annual Report to the Commander-in-Chief, U.S. Fleet, 1 July 1924 to 30 June, 1925, Admiral Robert E. Coontz, 62.

49. Carl E. McDowell, Helen M. Gibbs, *Ocean Transportation* (New York: McGraw-Hill, 1954), 256.

50. NARA, RG 19, Bureau of Construction and Repair, General Correspondence, 31, January, 1933, 2.

51. U.S. Naval Administration in World War II, Deputy Chief of Naval Operations (Air) Vol. 17, Procurement of Naval Aircraft, 1907-1939, 1946, 296. Government aviation employment exceeded commercial aviation employment.

52. U.S. Naval Administration in World War II, chief of Naval Operations, Vol. 17, "Procurement of Naval Aircraft, 1907-1939," Navy Department Library, 1946, 205.

53. Jacob A. Vander Muelen, *The Politics of Aircraft: Building an American Industry* (Lawrence: University Press of America, 1991): 55.

54. Hearings of the General Board of the Navy, U.S. Navy War College Library, Microfilm, July 1930, 262.

55. Oral History, Columbia University Library, Admiral Samuel M. Robinson, "The development of a light high speed diesel engine was very dear to my heart."

56. Gerald Gunderson, *A New Economic History of America* (New York: McGraw Hill, 1976): 33.

57. Douglas A. Irwin, *The Smoot-Hawley Tariff and the Great Depression*, draft manuscript, April 5, 2010. According to Irwin "...The legislative tariff increase was much smaller than commonly imagined. ...it was the deflation of prices that accompanied the Great Depression that pushed the tariff record levels."

58. The Davis-Bacon Act passed in 1931, required government contracting to pay the highest wage prevailing in a specific geographical area. The purpose of the act was to prevent wage competition.

59. Samuel E. Morison, Henry Steele Commanger, William Leuchtenburg, *The Growth of the American Republic*, (New York: Oxford University Press, 1980), 424.

60. Robert Oakes, *Battle Fronts of Industry*, (New York: J. Wiley, 1948): 49.

61. Louis M. Hacker, *The Course of American Economic Growth and Development*, (New York: John Wiley, 1970), 279.

62. 279.

63. L.S. Howeth, *History of Communications: Electronics in the United States Navy*, (Washington: Bureau of Ships and Office of Naval History, 1963), 368. Denby approved of the young plan in 1921 "...provided that a monopoly of transpacific communications would not be established in the United States."

64. Gleason L. Archer, *History of Radio to 1926*, (New York: American Historical Society, 1938), 77. Tony Devereux, *Messages, Gods of Battle: Radio, Radar Sonar, the Story of Electronics in War* (London: Brassey's, 1991): 75. "In the 20 years between the two world wars, civilian radio gave rise to mass public broadcasting... and produced television," also Susan J. Douglas, *Inventing American Broadcasting, 1899-1922* (Johns Hopkins University Press, 1987), 281.

65. Harless D. Wagner, *The U.S. Machine Tool Industry from 1900 to 1950* (Cambridge: MIT Press, 1968), 135-136.

66. Harvey S. Ford, "Walter Folger Brown, *Northwest Ohio Quarterly*, 1, 26, (Summer, 1954), 30.

67. Victor D. Seely, "Boeing Pace Setting 247," *American Aviation Historical Society Journal*, 9, 4 (Winter 1964), 239.

68. Henry Ladd Smith, Airways, *The History of Commercial Aviation in the United States* (New York: Alfred A. Knopf, 1942), 156-159. "Technology of War," 29, *The New Encyclopedia Britannica*, 1998, 64.

69. Thomas Kessner, *The Flight of the Century: Charles Lindberg and the Rise of American Aviation* (New York: Oxford University Press, 2010), 172-3.

70. John Braeman, "Power and Diplomacy: The 1920's Reappraised," *The Review of Politics*, 44, 3, (July, 1982): 345.

71. U.S. Congress, House of Representatives, War Policies Commission, 72nd Cong. 1st Sess., December 10, 1921, 354 [cited as war policies commission], 368.

72. 479-480.

73. Harold W. Thatcher, *Planning for Industrial Mobilization, 1920-1940* (Washington: General Administrative Service Division; Office of the Quartermaster Corps, 1943).

74. Lawrence J. Legere, *Unification of the Armed Forces* (New York, Garland, 1988), 155.

Chapter Four

The Franklin Roosevelt Administration, 1933-1941

Assuming office in early 1933, the Roosevelt administration found itself beleaguered by the economy's chronic unemployment. In the late 30's as Europe inched toward war, Congress and FDR began to turn their attention to the international scene, events that affected the policy matrix of the Navy Department.

ADMINISTRATIVE OVERSIGHT

Following Franklin D. Roosevelt's overwhelming defeat of Herbert Hoover in 1932, the new President appointed Claude Swanson as his Secretary of the Navy. Swanson, an attorney from Virginia, had chaired the Senate's Naval Affairs Committee for a number of years and was familiar with the navy's organizational structure and decision-making process.

Swanson was seventy-one when he accepted the Navy post. His health was less than robust. He suffered from pleurisy and later experienced a stroke that would confine him to a wheelchair. At one time, Swanson was absent from his office for the better part of a year. He remained Secretary until his death in July 1939.

Henry Latrobe Roosevelt, Assistant Secretary of the navy, an Annapolis graduate, had served as an officer in the Marine Corps. Henry Roosevelt also had health issues. In office less than four years, he was felled by a heart attack in 1936. President would later appoint Charles Edison as the Navy's Assistant Secretary.

Franklin Roosevelt's nearly eight years as Assistant Secretary during the Wilson era had given him an inside view of the naval establishment, its organization, tradition and personnel. As President, Roosevelt knew many officers on a first name basis, several of whom had now attained flag rank.

The health issues of Swanson and Henry Roosevelt did not impose any particular burden upon the President. Admiral William Standley, Chief of Naval Operations from 1933 to 1937, recalled that the President preferred to be his own Secretary anyway. According to one report, "The president kept a close check on who was buying the ship, what the ship would be used for and where the ship would be scraped." The President encouraged the bureau chiefs to visit him in the White House. Though ostensibly a meeting of former colleagues, the move tended to weaken the Secretary of the Navy and buttress the authority of the department's bureau chiefs. [1]

President Roosevelt had an abiding interest in the service. He named war ships, perused promotions lists, discussed naval strategy, thrived on service gossip and possessed a retentive memory—a gift that could work for or against a particular officer up for promotion. Franklin Roosevelt apparently had a strained relationship with Admiral Robert Coontz. The President refused to name a ship after the former CNO.

The administrative structure of the Navy surfaced in the early months of Roosevelt's presidency. Representative Carl Vinson, Chair of the House Naval Affairs Committee, concluded that the Navy's bureau chiefs possessed excessive power and autonomy. Vinson introduced legislation giving the CNO coordinating authority over the navy's bureau system. The chairman, however, was unable to pry the legislation out of his committee, much less bring the issue before the full House.

A second opportunity occurred when Admiral William Standley succeeded Admiral William Pratt as Chief of Naval Operations. A former member of the war plans division, Standley attempted to reinstate General Order No. 433, an order that gave his office coordinating authority over ship repair and alteration. Rear Admiral William Leahy, Chief of Ordnance and Rear Admiral Ernest King, Chief of Aeronautics apposed the order. Secretary Swanson asked Assistant Secretary Henry L. Roosevelt to form a committee to look into the matter. The committee rejected Standley's General Order 433, a decision reaffirmed by the Assistant and Secretary of the Navy. Standley took his appeal to the President. Roosevelt responded that CNO coordination authority did not give the CNO Navy bureau oversight and he vetoed Admiral Standley's administrative move. The President, however, refused to cancel the General Order outright. He simply stated that it did not apply to the Office of Naval Operations. [2]

A third attempt to integrate Operations and the Bureaus experienced a similar fate. Assistant Secretary Henry L. Roosevelt proposed to set up an Operations Council made up of the Chief of Naval Operations, the department's budget officer, the bureau chiefs, the Marine Corps and the Judge Advocate General's office. The group planned to meet with the Chief of

Naval Operations to iron out war planning problems. President Roosevelt axed the proposal on grounds that an Operation's council had the trappings of a formal staff system. Claude Swanson supported the President.

One incident highlighted the issue of bureau coordination in the late 1930's. The Navy had completed the construction of the destroyer Anderson, as a prelude to standardizing the production of additional destroyers. Built by the Federal Shipbuilding and Dry-Dock Corporation, the Anderson, loaded with fuel and munitions, performed well in a trial test. The next step was to see how the destroyer handled with low fuel on board. The Anderson demonstrated instability, a problem that reignited the issue of bureau management and coordination.[3]

Three material Bureaus, Construction and Repair (hull), the Engineering (machinery), and the Ordnance (armament) were responsible for a section of the ship's overall design. The Anderson's top heavy problem suggested that bureau cooperation was less than optimal. At first the Bureau's disclaimed any responsibility for the destroyer's problems, insisting that a New York design firm, Gibbs and Cox, was the real culprit. Gibbs and Cox researched their correspondence archive and pulled out letters informing the Navy that the Anderson was top heavy. The ball was back in the Navy's court.[4]

After the passing of Henry Latrobe Roosevelt, President Roosevelt appointed Charles Edison, the son of the famous inventor, to the post of Assistant Secretary. By virtue of running his father's manufacturing enterprise, Edison brought to his office a modicum of business experience. Edison also possessed a discerning, if not clinical, eye. After taking office, Edison concluded that the Navy's operating fleet was in fairly decent shape. He was less than impressed with the navy's shore establishment. The design flaw of the Anderson confirmed Edison's skepticism.

In a move toward departmental coordination, Edison proposed that engineering, construction and repair, and ordnance form a liaison group as a vehicle to address future design issues. Edison asked that Congress to consolidate the three bureaus into a single, unified Bureau. Not only did the Bureau of Ordnance object, but President Roosevelt himself poured cold water on the idea. Subsequently, Congress passed legislation that folded both engineering and construction into a new Bureau: the Bureau of Ships. Not only did the President approve the consolidation, but he added a new civilian post, an Under Secretary of the Navy. Presumably that office would ease the administrative burden of the Secretary.[5]

Still, FDR did not appear particularly concerned by the Navy's administrative tensions. The service was thought to possess a deep bench of experienced officers. When Secretary Swanson was absent from his office due to travel plans, illness, or conferences, the Assistant Secretary could step in as acting Secretary. If that were not possible, the CNO could take over.

In late 1936, Secretary Swanson was ill, Assistant Secretary Henry La-trobe Roosevelt had died and Admiral Standley attended a London naval disarmament conference. FDR felt that Charles Edison, given his shore es-tablishment responsibilities, had more than enough to do. Accordingly, the President designated CNO Admiral William Leahy as Acting Navy Secre-tary, an assignment that tended to erode clear demarcations between civilian and military authority. Admiral Leahy routed a memo to the Chief of Naval Operations recommending a ten year construction program. Admiral Leahy signed the memo "acting" Secretary and sent it to the Chief of Naval Opera-tion. Admiral Leahy, in effect, sent a memo to himself.[6]

Leahy's ambivalent position did not appear out of place. The President had known William Leahy as a naval officer as far back as 1915 and held him in high regard. In any event, someone in the Navy had to move the department's administrative paper. When Leahy took retirement in 1939, the President gave serious consideration to appointing the former Navy head as Swanson's successor. Instead, Roosevelt selected Frank Knox, a newspaper publisher and a Republican. Unlike Woodrow Wilson, Roosevelt decided it was time to reach across the political aisle.

BALANCED FLEET

Although the 67th Congress and 71st Congress approved the Washington and London naval agreements affirming parity between Britain and the U.S., the House and the Senate refused to appropriate funds to bring the U.S. fleet to its allowable treaty strength. The great depression intervened and the fall-off in tax receipts compelled the government to cut spending across the board.

The Navy had anticipated that FDR's election victory might rectify the fleet's tonnage short fall. They were not disappointed. Congress passed the National Industrial Recovery Act (NIRA) of (1933), as a job creation pro-gram. By executive order, the President transferred $280 million dollars from the NRA to the U.S. Navy—bypassing, in the process, congress's legislative authority.

The next year, 1934, Congress passed the Vinson-Trammel Act to bring the U.S. fleet up its assigned treaty level. Via executive order, FDR trans-ferred $28 million dollars from the Works Progress Administration to the Navy, this time circumventing the vetting authority of the Budget Bureau. The President was obviously in a hurry. There were secondary repercussions, nevertheless. William M. McBride concluded that Roosevelt's $280 million dollar budget, "...stimulated naval expansion in Japan." Parties to the naval treaty system monitored each other's actions with utmost vigilance.[7]

The Japanese government, still smarting because their fleet had been denied tonnage parity, announced in 1934 their intent to withdraw from the treaty system. Three years later, Japan was at liberty to embark on a warship program of its own choosing. Nor did the U.S. mark time. Congress added seventeen new battleships to the fleet's inventory of sixteen battleships, a program supported by CNO Admiral William D. Leahy, (1937-1939) and his successor Admiral Harold R. Stark (1939-1942). U.S. lawmakers did question why Admiral Leahy as to why he excluded additional aircraft carriers from his proposed budget. The CNO responded that the navy's flying boats could provide the fleet's scouting assignment. [8]

Although military aircraft experienced technical advances during the 1930's, some seventy five percent of U.S. military planes were either biplanes or underpowered monoplanes. Commercial aviation, incorporating improvements in speed, range and carrying capacity, began to eclipse the capability of military aircraft. [9]

The fleet's command structure invited debate as well. Secretary Swanson turned down a General Board's proposal to adopt a task force command protocol. Admiral William Standley, on the other hand, supported a matrix command structure. In choosing Rear Admiral Joseph Reeves as CINCUS, Admiral Standley appointed an officer not only wedded to a type command format, but resisted Standley's attempt to exercise authority over the fleet's Commander-in-Chief. [10]

The General Board addressed the issue of fleet command in the early 30's. In attendance were Admiral W.H. Standley, Chief of Naval Operations; Admiral J.M. Reeves, commander-in-chief, U.S. Fleet; Admiral J.M. Greenslade, a member of the General Board. A selected colloque gives some sense of the discussion.

Admiral Reeves: "The Battle Force and Scouting Force is, I believe, a premature war task force organization before we know what the war is about."

Admiral Standley: "If you have no permanent task group... would not the Commander-in-Chief have to initiate every coordinated effort?"

Admiral Reeves: "Not any more than we do today."

Admiral Standley: "Has the Commander-in-Chief got time to do that detail work himself and take care of the problems which they expect him to do?"

Admiral Reeves: "He would have more time than he has with the present organizations."

Admiral Greenslade: "....all the results of our war college work have brought us to a common school of thought which is what we call the task force operation."

Admiral Reeves: "....we have avoided the fatal mistake of creating a task group to meet an unknown situation that it might not fit except by chance." [11]

Obviously, Admirals Reeves and Standley held divergent views on the matter of fleet command. From the perspective of congress, the Navy's command structure appeared to be a matter of internal, parochial interest. The House and Senate preferred to devote their attention to naval bases and stations.

The fleet's annual exercises continued as a platform to test tactical doctrine and force structure. The battleship remained the fleet's premier weapon, the carrier, submarine, destroyer and cruiser defined by their battleship support. Although the speed and mobility of the carrier suggested that its potential was not confined to the role of a combatant auxiliary, the doctrine of the fleet appeared over-riding. The battleship remained the fleet's backbone.

A tactical consensus of sorts began to emerge as a result of annual fleet training problems throughout the 1930's. They included the following:

• Carriers and surface warships were now integrated into the fleet
• Carrier speed signaled an offensive capability all of its own
• The concept of a carrier task force appeared workable
• Catapult planes on warships were damaged by gunfire vibration
• Carrier planes that struck first appeared to prevailed over enemy carriers
• Angle dive bombing increased bombing accuracy
• Airships seemed vulnerable as fleet assets

In the meantime, the fleet and Naval Research Laboratory collaborated in developing short wave radio communications. Scientists at the Navy Research Laboratory (NRL) had engaged in radio detection experiments in 1922 and again in 1930. In an era of tight budgets, NRL scientists, on occasion, resorted to sub-rosa research funding. During off hours, they built detection equipment, buying apparatus on their own, occasionally poaching on resources from other research projects.

Corporate research laboratories remained part of the industrial scene in the 1930's, although General Electric, Westinghouse, and RCA cut back expenditures. Amateur radio enthusiasts demanded high performance vacuum tubes in their quest for long distance communications. Viewing standard of the shelf radio tubes as inadequate, amateurs turned to be such specialists as Heintz and Kaufman and Eitel McCullough for high performance tubes.

Custom made tubes now occupied a market niche. One radio historian noted that, "the amateurs-the unknown soldiers of science who explored the desert regions of the spectrum and gave to the world high frequency, super-high frequency and micro-wave radio."[12]

In 1934 the NRL hired Robert Page to work on radio research activities. Page assembled a detection prototype that demonstrated some technical promise. Still, NRL budgets were tight, and although the laboratory was now housed in the Bureau of Engineering, the Bureau provided little funding for high frequency radio.

NRL scientists approached Representative James Scrugham, a member of the House Naval Affairs Subcommittee. The researchers outlined their radio detection work. Congressman Scrugham listened and said little. The NRL scientists returned to their laboratory less than elated. The next day, the NRL learned that James Scrugham had inserted $100,000 for radio experimentation. Scrugham was not trained in the law. He was, by profession, an engineer. [13]

A year later, Page came up with a duplexor that enabled two transmitters and receivers to share a single antenna. Page found that off the shelf radio tubes could not generate short waves at high frequencies and low wattage. Page turned to a small short wave tube supplier, Eitel-McCullough. The company's *Eimac* generated a wave length of 200 meters/sec.

The next year, 1937, the NRL set up a radio detection prototype on a World War I destroyer. The set picked up the echo of a ship some 20 miles away. In 1938, the Commander in Chief (CINCUS) divided the fleet into offensive and defensive units. By this time the NRL invited RCA to participate in detection research work. RCA placed its prototype on the *U.S.S. Texas (XAF)*; NRL's prototype on the *U.S.S. New York (CXZ)*. The *New York* prototype picked up echo of a plane 100 miles out. Interestingly, McCullough had shipped their tubes to the NRL by air freight.[14]

Naval officers participating in fleet exercises, soon became radar converts. In his report, Admiral A.W. Johnson wrote that the *New York* radar set was, "...one of the most important military developments since the advent of radio." *New York*'s Captain R.M. Griffin recommended that the radar prototype, "...be installed at once on all [aircraft carriers] and as soon as possible on other vessels." The Navy next entertained production bids. RCA beat AT&T's Western Electric for radar sets although Bell Labs and Western would receive a contract to supply gun control radar apparatus. [15]

The U.S., of course, was not the only contender in the radio identification field; France, Germany, Italy and Russia were active as well. It was Britain, however, that set the pace. Concerned by Germany's aviation built up in the 1930's, British scientists and military set in motion the construction of a radar early warning system. By the time Britain's expeditionary forces had been evacuated from Dunkirk in June 1940, Britain had constructed 27 detection stations scattered along the Island's east coast.

Britain next assembled a radar/communication network that fed enemy plane detection signals into a centralized command center. The network permitted the RAF to identify actual enemy planes rather ordering RAF pilots to patrol the skies all day long. Fatigue under the stress of combat is often as a critical variable.

Dr. Henry Tizard, cleared by Prime Minister Winston Churchill, led a technical mission to the U.S. in August 1940. Britain unveiled its military secrets to the U.S., including plans for the Rolls Royce engine, a proximity fuse, and jet engine development. Tizard's cargo included a pulse cavity magnetron. Developed by Birmingham University physicists, the magnetron generated a wave length of 10 centimeters/sec. The magnetron would lay the foundation for microwave radar. U.S. naval research scientists immediately recognized the potential of centimeter radar. [16]

Tizard asked that the U.S. radio and electronic industry manufacture and supply the next generation of radar. The U.S. agreed to do so. Adopting Britain's science research model, the U.S., through MIT's Radiation Laboratory would later design over 150 radar systems. [17]

AMPHIBIOUS OPERATIONS

The Marine Corps survived a potential Army takeover during the Hoover administration. Perhaps aware of that close call, the Marines were determined to align their destiny with the Commander in Chief, U.S. Fleet. In 1934, the Marine Expeditionary Force, renamed the Fleet Marine Force (FMF), was designated to serve under the command of the U.S. Fleet (CINCUS).

The U.S. economy continued to falter in the early 1930's. Low on funds, the Marines called off their landing exercises and embarked upon a project to consolidate what the Marines had learned about amphibious warfare. A remarkable document emerged from that endeavor—a tentative plan that laid out the principles, guidelines, and mission of amphibious or joint operations.

Between 1935 to 1939, the Marines participated in fleet problems and landing exercises—the latter known as Flexes. On occasion, Army troops joined in the exercises. Each exercise generated its own post mortem; and the refrain became repetitiously familiar. The Marines lacked transports, cargo ships, field artillery, tank lighters, landing craft, aircraft, and battleship gun support. Although radio sets continued to be temperamental and unreliable, the Marines regarded radio as imperative for unit coordination. The Marines, in fact, asked Pan American Airlines if they could purchase their RCA radio apparatus. RCA turned down the request.

It was also conceivable that the Marines were not essential to the Navy's Pacific war plan. An orthodox plan called upon the Battle Fleet to sortie to the western Pacific where it would meet, engage and destroy the Imperial Japanese fleet. Assigning a warship to amphibious gun support detracted from the battle fleet's prime target, the destruction of the enemy's ship of the line.

Nor did the Marine's frustration end there. On occasion, some Naval officers could not resist denigrating the Marines or resorting to the corps as a form of punishment. More important, amphibious operations implied non-combat duty, a professional kiss of death. As one aphorism had it, an officer might be willing to sacrifice his life for his country, but not his career. [18]

The Marines' request for close ground met a tepid response from Naval aviators. Carrier pilots were trained to protect the battle fleet, to serve as fleet reconnaissance, to search and destroy enemy aircraft and ships. Close ground support reduced the standing, not to say the prestige of naval fighter pilots.

Throughout the 1930's the Marines wrestled with nonexistent transports, aged cargo ships, ill-designed landing craft. Driven to improvise, the Marines discovered World War I destroyers mothballed in storage. By removing a boiler, the destroyer could accommodate 100 assault troops. The hand me down destroyers, classified as fast amphibious transports (ADP), permitted the Marines to engage in raiding strikes.

By the mid-thirties, Germany began testing the boundaries imposed by the Versailles Treaty. Two years later, China found itself under Japanese siege. General Malin Craig, army chief of staff, asked Admiral William Leahy, CNO, if infantry troops could participate in joint marine/navy fleet maneuvers. Admiral Leahy rejected the request on grounds that the landing exercises were preliminary and need not involve infantry participation. [19]

Three years later, U.S. Army Brigadier General James G. Ord, after his experience in landing exercises, wrote the following final report;

> The plans of the fleet commander (Admiral Ernest King) to train his fleet in maneuvers preparatory to landing caused him to move the fleet to sea each day following a landing. This required the re-embarkation of troops and impedi-mendia to start by daylight of the morning after a landing. I judge from the general requirements of naval training that the situation will continue in future maneuvers. I believe the training in this part of the landing is so important that it should be arranged for in the future. It can be done without fleet participa-tion. [20]

General Ord's memo served as a prelude to the Army's decision to establish its own amphibious unit, an Engineer's Combat Battalion. By 1941, the Army and Navy appeared to go their separate way on joint operations. To some naval officers, that was well and good. In any future conflict, the Army could concentrate on European Theater, the Navy and Marines deal with the

Pacific Theater. The matter would remain unresolved until the Second World War.

In 1939, Admiral C.C. Bloch, Commander-in-Chief, U.S, Fleet, submitted his annual report that summarized his reaction to fleet problem #20. His critique appeared to be a wakeup call.[21]

a. Lack of docking facilities. The only floating drydock operated by the Navy and which could be made available at an advanced base is the ARD-ONE which is limited by its size to docking of 1500-ton destroyers. Lack of docks will severely handicap operations.

b. Lack of distilling ships. The lack of distilling shops may bring about a critical situation at an advanced base as many of the ships which will be taken over by the navy lack evaporator and distiller capacity and depend upon fresh water storage tanks, which, while sufficient for short runs, would not suffice for the service which would have to be met under war conditions.

c. Lack of fast mine sweepers. This condition is being remedied as soon as practicable by the use of old destroyers. Mine sweeping gear is, however, still in the experimental stage. It is hoped that experimentation during the coming year will solve this problem and that it will be possible to equip sufficient old destroyers with standard gear in case of emergency.

d. Lack of submarine nets and net laying vessels.

e. Lack of suitable landing boats.

f. Lack of equipment for rapid construction of landing fields.

g. Lack of suitable store ships.

h. Lack of high speed oilers.

i. Lack of antiaircraft guns and fire control equipment for auxiliaries.

j. Fleet Landing Exercise No. 5 again demonstrated the need for artillery lighters of suitable weight to be handled by the cranes or booms on most naval or commercial vessels. It is hoped that new methods of construction and the use of lighter materials may provide an early solution to this problem, and that a sufficient number of such lighters will be provided at designated loading points to meet the early fleet requirements in even of emergency.[22]

Admiral Bloch also remained committed to the battleship. When Captain Frederick Horne distributed a post exercise paper suggesting the carrier's aviation role in future fleet action, Admiral Bloch ordered Horne to recall his paper. Horne did more than that. After collecting his report he burned every sheet.[23]

The Fleet Marine Force, by this time, had codified its experience in a landing operations manual. The manual included the role of command, battleship support, aviation assistance, beach parties and logistics. In its Flex exercises, the Marines refined and honed their tactical doctrine. By 1937, the navy's Fleet Training Division published F.T.P. 176. The Army later issued a manual 31-5. The two were essentially identical.[24]

LOGISTICS

In January, 1937, Japanese planes sank both the U.S. gunship Panay and a Standard Oil Company oil tanker (New Jersey) on the Yangtze River. The Navy directed Admiral Arthur Hepburn to review U.S. base needs. Hepburn came up with an eye-popping $287 million dollar investment number—including an upgrade of Guam. Concerned that bulking up Guam's defense might antagonize Japan, U.S. lawmakers excised the Pacific Island's fortifications from the Navy's base request.

The status of Naval oil reserves no longer dominated the nation's newspapers in the 1930's. Reserve #1, remained semi productive; reserve #2 drained by adjacent private drilling; reserve #3 Teapot Dome, had been capped; oil reserve #4 remained unexplored. Salt Creek, the non-navy oil reserve in Wyoming, continued its productive output.

The Navy had placed oil storage depots on both U.S. coasts, Panama, Pearl Harbor and Guantanamo. The Pan American Oil Company had built a pipeline feeder system in California. Sinclair' Mammoth Oil Company constructed a 900 mile pipeline that tied Wyoming's Salt Creek to pipeline facilities in East Chicago, permitting oil transfer to Northeast and Gulf coast.

The petroleum industry experienced consolidation in the late 20's and 30's. Standard Oil of Indiana acquired Doheny's Pan American's Lake Maracaibo operations in Venezuela. After Congress imposed a tariff on petroleum's imports, Indiana Standard, incurring an operating loss, sold its Maracaibo holdings and Aruba refinery to Standard Oil of New Jersey. The acquisition also included Doheny's Maracaibo drilling wells and San Lorenzo refinery. Jersey Standard employed shallow drafted tankers designed to convey Maracaibo crude to its Aruba refinery.[25]

The U.S. Fleet continued to rely on ship forces for repair and maintenance afloat. The fleet also engaged in refueling exercises. By the late 1930's, the navy's CNO (Stark) activated fleet oil replenishing with renewed urgency. At the time, the Navy had some 29 fleet oilers. In the event of war, the Navy viewed merchant tankers into fleet auxiliaries. Admiral Harold stark ordered the fleet to conduct refueling operations between Esso tankers and U.S. warships.[26]

In 1935, President Roosevelt asked Secretary of State Cordell Hull whether the U.S. should legislate a merchant ship building program. Hull, who fourteen years earlier had violently celebrated the defeat of Harding's ship subsidy plan, responded with a resounding yes. Congress passed the U.S. Merchant Marine Act of 1936, legislation that contemplated two financial stipends – one for private ship yards development, the other for fleet operators based on vessel speed. The U.S. Shipping Board, renamed the U.S. Maritime Commission, adopted a Standard Oil (New Jersey) tanker design, capable of 18.5 knot speed. Standard Oil in turn agreed to purchase and add twelve tankers to its commercial fleet. The head of Standard's tank fleet was a naval reservist.[27]

The Navy belatedly recognized that refueling at sea enhanced the fleet's operating range and mobility. The fleet also attempted to stretch the time warship before a ship was due for a dry dock visit. Hence, the Navy continued research on antifouling paint. By 1940, the Mare Island Navy Yard had developed a hot plastic paint, 15HP, and a cold plastic paint (143E).[28]

SOURCING

Make

From whom the Navy secured its material requirements boiled down to the usual choices; rely upon the government arsenals; purchase from the nation's industrial sector. Congress legislated the Vinson-Trammel Act, together with Navy appropriations in 1933. The Navy began to build up its fleet inventory. Congress then did the unexpected. Rather than assigning a ship construction to a particular Navy Yard, Congress delegated that decision to the President. When the President set aside $280 million dollars of NRA money for ship construction, he did so on the assumption that labor constituted 80% of a vessel's construction cost.

U.S. lawmakers insisted that government yards provided a benchmark to assess private ship yards and their construction costs. That yardstick rested on an unspoken premise that government yards had knowledge of their own costs. That Franklin Roosevelt manifested some doubt is suggested by a memo he sent to the Navy Secretary in 1940 asking, why government ship's cost so much - and why they were delivered late.[29]

A related study addressed the status and condition of government plants and arsenals, concluding that "…85% of its (government) machinery was ten years old, much of its antedated the turn of the century; some had actually been installed before the civil war." A later ordnance study described the culture of union work rules at the navy's torpedo station, Newport, Rhode Island. The study observed that;

> …Torpedoes could be built only by craftsmen who knew the proper trade secrets. These secrets, instead of being properly committed to writing or draw-

ing and in specifications, were largely matters of memory on notes in some foreman's little black book at Newport station.

A few naval officers questioned the efficiency of the department's yards. Congress, nevertheless, insisted that government yards existed to keep private yards honest.[30]

There is little evidence that 74th-75th Congressional staff members audited government yards; questioned whether vessel components were delivered on time; probed whether ship construction experienced cost overruns; asked whether deficiency budgets masked yard cost over-runs; determined whether yard employees operated modern machine tools; explored whether political pressure compromised yard productivity; assessed the cost of trade union work rules. Such matters were generally assumed away. It was enough to know that government yards created jobs, that yards generated a payroll, that yards provided patronage, that yards delivered votes. Some legislators may have suspected that the navy's shore establishment may not have been the most inspired model of productivity given the ban on Taylor practices. At least, government plants removed the taint of corporate earnings.

Outsourcing

Nor did the 73rd-75th Congress's deem it of imperative to question whether the Bureau of Construction and Repair wore conflicting procurement hats. The Bureau designed ship hulls before sending construction plans to a Navy Yard. The bureau also acted as a buying agent on behalf of the operating fleet. Put differently, the Bureau evaluated its own product as well as that of its rivals. That an outside supplier might find itself foreclosed from a government holding company vesting an interest in protecting its own supply affiliate was deemed beyond the threshold of congressional interest. Sustained by congressional passivity, the Navy's vertical structure remained in place, despite occasional questions raised by Representative Carl Vinson.

The Marine Corps, by contrast, continued to ask the Bureau of Construction and Repair to design a personnel landing craft. The bureau moved ponderously. The Bureau's response appeared to be a classic case of bureaucratic stall. In frustration the Marines set up their own Landing Craft Committee and began looking beyond the Navy's shore establishment for its amphibious needs.[31]

In that process, the Marines found themselves intrigued by a craft designed by Andrew Jackson Higgins. Admiral William Leahy, CNO, however, assigned a Bureau designed craft to be built at the Philadelphia and Norfolk Navy Yards. When the Marines questioned that production rationale the CNO responded that Navy Yard employees needed work. One historian concluded that when the navy's vertical structure trumped the needs of the

Marine Corps it constituted "…an indefensible bureaucratic decision to purchase five boats simply because they were designed by the BCR." (Bureau of Construction and Repair) [32]

The 74th-75th Congress directed their attention to the structure and organization of the nation's industrial sector – still under financial stress. Concluding that the cause of the nation's chronic unemployment resided with the nation's private sector, Congress passed legislation that broke up banking holding companies, restructured electric power holding companies, funded a government holding company of its own, the Tennessee Valley Authority. As a device to increase jobs and personal income, Congress legislated collective bargaining agreements supplemented by minimum wage and hour regulation. To cover anticipated spending deficits congress raised taxes on personal income, corporate earnings and undistributed corporate profits. By 1940, the personal income marginal rates approached 90%.

The economy failed to respond to New Deal policies. By the late 1930's the nation's shipbuilding yards now experienced a cost-price squeeze. On one side, outside rivals held down construction bids; on the other side, federal mandated hours, minimum wages, prevailing wages and overtime, elevated construction costs. Government yards continued to post estimated prices while private yards required to submit firm prices. Invariably government years under bid their private rival. After FDR awarded ship construction to a Navy Yard, the President, not unlike his mentor, admonished private builders to "sharpen their pencils." [33]

By January 1, 1937, the naval treaty system had expired, releasing the U.S. fleet from any tonnage constraint. FDR proceeded to assign battleship construction to three large eastern government yards, New York, Philadelphia, and Norfolk. Labor intensive and steel hungry, each battleship required 30 million hours of labor work. In funding the battleship program, Congress appeared intent on reviving Wilson's 1916 naval legislation. Private yards, of course, accepted contracts to build cruisers, carriers, and destroyers. But the President, assigning battleship construction largely to the government yards, engaged in a classic case of tonnage cream-skimming, aided and abetted by trade union employees. As a bonus, Congress prohibited Navy Yards from adopting modern management practices-Taylorism. Indeed, the International Machinist Unions in 1935 attempted to prohibit yard welding training programs that competed with riveters. [34]

The 73rd-75th Congresses blanketed commercial shipbuilding yards with semi utility oversight. To qualify for Navy contracts, private firms had a demonstrate shipbuilding experience, not unlike a certificate of public convenience and necessity. To ensure compliance, Congress imposed a ceiling on corporate earnings, though the incidence of risk was somewhat biased. The Treasury Department shared corporate profits; equity owners bore capital losses.

The National Industrial Recovery Administration (NRA), the centerpiece of the New Deal, acted as a forum for bidder access. Accordingly, a Pensacola, Florida Company, Gulf Industries, Inc. sought to qualify as a government contractor. A southern consortium, Gulf Industries included an Alabama Steel Company headed by a former president of the Federal Shipbuilding and Dry-dock Company of Kearny, New Jersey. A start-up firm, Gulf Industries depended upon receiving a line of credit from the Reconstruction Finance Corporation (RFC).

East coast commercial shipyards united in opposition to Gulf's bid. Private yards insisted that experienced yards only could qualify for government work - a position backed by Chief of the Bureau of Construction and Repair, Rear Admiral Emory Land. After a round of hearings the NRA dropped Gulf Industries as a contender.[35]

The Board's rejection of Gulf Industries actually proved unnecessary. The Governor of New Jersey lobbied the RFC to reject Gulf Industries' loan application; and President Roosevelt, aware of the Gulf's precarious financial status, dismissed the company as a non-going concern. Under the guise of pursing the public interest, congress and the administration preceded to cartelize the nation's ship building industry.[36]

Admiral Hutchinson Cone, former Chief of the Bureau of Construction and Repair and a member of the U.S. Shipping Board, testified on behalf of Gulf. Despite the fact that the U.S. was in the throes of an economic depression and that excess capacity was very much an economic reality, Admiral Cone predicted that, within a decade, the U.S. would experience a shipyard shortfall. The former Bureau Chief added a somber note. The nation's shipbuilding capability was overly concentrated on the nation's east coast. The U.S., he said, should geographically diversify its construction awards.[37]

Congress was presented with an opportunity to spread its contracts beyond the U.S. east coast. Eastern commercial yards, located within striking distance of Pittsburgh's coal and steel supplies, enjoyed a cost advantage over west coast yards. West Coast yards imported ninety percent of their steel from the east coast factories. Burdened by a cost disadvantage, western representatives proposed to balance the playing field by according western yards a 6% cost differential for the purposes of government contract work. The 74th Congress joined Secretary Claude Swanson opposition to the 6% price differential. The U.S. Maritime Commission, by contrast, adopted the 6% surcharge in letting merchant ship contracts to the west coast. In hindsight, one could argue that the Maritime Commission exercised greater strategic perspicacity than that of the Navy Secretary.[38]

It was also difficult to square the Secretary's 6% decision with the Navy's War Plan Orange strategy. The Navy had long viewed Japan as a potential adversary, and private yard construction on the west coast supported with any planned trans-Pacific naval operation. The nation's ship yards remained, nevertheless, concentrated in the east. Absent local steel suppliers, West coast yards found themselves beholden to Pittsburgh plus steel pricing.

The navy's vertical structure did invite a critical review by Rear Admiral H.G. Bowen, Chief of the Bureau of Engineering. Bowen concluded that stationary power plant turbines were comparable to units sold by traditional maritime suppliers. Large integrated private yards, Bethlehem, New York Shipbuilding Corporation, Newport News preferred to build parts and components in-house. Independent shipyards, by contrast, subcontracted ship parts and components. Bowen became convinced that non maritime products were technically advanced over the navy's in-house components. In attempting to loosen the buying patterns of the navy's shore establishment, Bowen encountered institutional resistance. [39]

Charles Edison, Assistant Secretary of the Navy detected a not-invented-here attitude (NIH), among government shipyards. When the Assistant Secretary proposed to import a British P.T. boat, a southern Congressman threatened the Secretary with impeachment. A contract award to a U.S. boat supplier, apparently, assuaged the lawmaker and the congressional threat faded away. The lesson was unmistakable. In naval procurement matters, a Secretary had best exercise political prudence. [40]

Following the 1936 Merchant Marine Act, fast commercial tankers were beginning to narrow the speed gap with heavy warships. Rear Admiral Emory Land, former Chief of the Bureau of Construction and Repair and Maritime chairman, later reflected on the Navy's indifference to fleet oilers. The Navy, he recalled was overly wedded to heavy warships construction. [41]

Admiral Land engaged in a remarkable process of circumvention. He employed a civilian agency, the U.S. Maritime Administration, to bypass the Navy's integrated yard complex in order to secure merchant bottoms capable of conversion to naval auxiliaries. In so doing, Land side stepped a ban on scientific management (Taylorism). Emory Land had long accepted the proposition that any U.S. Naval engagement in the Pacific would be auxiliary dependent. Land, in short, understood the nexus between logistics and dual use assets.

Incentives

Congress remained baffled by the persistence of double digit unemployment in the 1930's. Congress improvised a range of experiments to alleviate what they perceived as casual factors. A first argued that the U.S. economy was beset by excessive market rivalry that reduced wages, earnings and consumer incomes. Market competition inhibited the economy's recovery. As a solution to insufficient jobs, Congress suspended the nation's antitrust laws and promulgated a nationwide price and wage fixing program. Under government sponsorship, former corporate rivals could now engage in price collusion. Companies that continued to discount prices were labeled chiselers; individuals who supported price control sported a blue eagle badge on their jacket lapel.

In a New Deal setback, the Supreme Court struck down Congress's price fixing legislation, a setback that incited Congress to throw down the gauntlet. Congress adopted a new explanation. The cause of the economy's anemic performance was not the presence of open, unfettered market rivalry, but rather its absence. Congress asserted that corporate America had caused the great depression. A few lawmakers took the proscription a further step. Business, they insisted, had also promoted the First World War. Ad hominems followed. Congress labeled firms as "merchants of death," "banksters," "money changers." Commerce became less than a term of endearment. The fact that the U.S. Congress voted to support Woodrow Wilson's war declaration was all but forgotten as lawmakers enumerated a bill of particulars against the nation's industrial and banking officers. [42]

Addressing economic concentration, Congress charged that large firms, refraining from engaging in traditional price competition, preferred, instead, to engage in non-price rivalry—notably, advertising. The Senate and House insisted that giant firms imposed administered prices upon an unsuspected, unprotected, helpless buying public.

With this diagnosis in hand, Congress activated a program of antitrust action. In 1938, President Roosevelt Budgeted $413,000 for antitrust action in funding. By 1942, the budget was scheduled to reach $2.3 million. The Justice Department filed restraint of trade and antimonopoly suits against U.S. firms as Congress raised taxes on personal income, corporate earnings, undistributed corporate earnings, inter-corporate dividends, gift, luxury and estate levies. All were attempts to transfer income from the affluent to the less fortunate - a device geared to ignite the nation's purchasing power. Antitrust thus joined monetary and fiscal policy as a macroeconomic tool. U.S. lawmakers, however, neglected to calculate the compliance cost that federal taxes, federal regulations imposed on the private section. In the discourse that followed Congress remained perplexed that U.S. net private investment between 1930 and 1940 remained "...a negative $3 billion dollars." [43]

The Senate and the House next dusted off Woodrow Wilson's Muscle Shoals program. Renamed the Tennessee Valley Authority, (TVA) the TVA operated hydroelectric dams and power to residential and commercial customers. Cheap electric rates would ensure economic growth and development in the south.

Few lawmakers questioned the cost of providing "cheap" government electricity. The fact that a government agency employed tax free assets to compete with its private electric rivals; that the Public Works Administration lent money to TVA for the acquisition of electric municipal customers; that

TVA's electric rates failed to cover relevant foregone costs; that government enjoyed lower interest payments than those borne by private firms; the fact that TVA's electric cost excluded capital equipment tied to "conservation"— all such accounting manipulations were justified by a pursuit of the public interest. President Roosevelt, in fact, was so taken by the TVA model that he wanted to build six more hydro-electric dams throughout the U.S. [44]

Nor could Congress resist micromanaging the nation's private sector. Turning his attention to the Post Office's airmail subsidy program, Senator Hugo Black of Alabama, charged that Hoover had cartelized the air transport industry. Black recommended that the government nationalize commercial aviation. Instead, President Roosevelt approved Postmaster General James Farley's decision to cancel the mail subsidy program. Farley then ordered the U.S. Army Air Corps to deliver the nation's airmail. The death of 12 of army pilots within two weeks precipitated an administration crisis. The Post Office also discovered that a subsidized Army Air Corps letter escalated from $.57 to $2.50.

As part of its pilot re-training program, the Air Corps purchased flight simulators manufactured by Edward Link. President Roosevelt ended the Air Corps' experiment and discretely reinstated President Hoover's original air route system - but not before exacting a price. To qualify for government mail subsidies aviation holding companies had to spin off their plane and engine supply affiliates. [45]

Congress continued to enhance the economic power of organized labor. Wilson had immunized unions from the reach of the nation's anti-trust laws, exempting a union from trade restraint suits and damage suits. Congress passed the Wagner Act that outlawed unfair labor practice and set up a board to adjudicate union elections. Using a congressional committee as its operating model, the National Labor Relations Board served as judge, jury and prosecutor in adjudicating labor and management disputes. Congress then legislated the Fair Labor Standards Act, a bill that promulgated minimum wages to workers in the nation's private sector. Congress, in short, legislated price inflation.

The Treasury Department continued to pursue its own menu of economic disincentives. The department imposed an amortization life of twenty years for tax purpose, a write off that reduced expenses and lifted corporate earnings into a higher tax bracket. A five-year write off, by contrast, raised annual depreciation expense, reduced earnings and cut a firm's tax liability. As a vehicle to enhance government tax revenues, the Treasury Department preferred the former.

The air frame industry experienced its own set of economic disincentives dating back to competitive bidding rule of 1926. Under the act, Congress socialized private aviation prototypes submitted for government contract purposes. If the government reimbursed a company for its prototype, the cost often fell short of the firm's development cost. An inverted scenario followed; private research expenditures subsidized government aviation awards.[46]

In a sense, government officials acted as if they were dealing with a commodity - wheat or barley. In reality Navy buyers were purchasing a technology that, in a matter of months, could be superseded by a more advanced design. Unmoved, Congress married a static mindset to an environment of product dynamism. Any lawmaker who suggested that government policy stifled rather than enhanced the nation's aviation industry risked being vilified as being in the pockets of the nation's munitions makers.[47]

The passage of the 1934 Vinson-Trammel Act (1934) to bring the Navy to its treaty limit was touted as a pro defense measure. Congress proceeded to treat plane manufacturers as if they were local water companies. The law imposed an 8% ceiling on contractors' profits. Vinson-Trammel, however, applied to the Navy, not to the War Department. Noting the discrepancy, the Boeing Company dropped its Navy contracts and concentrated on building Army bombers. The Boeing carrier fighter XF4B that had given a fine account of itself against Army fighters during the 1929 fleet exercises was now phased out. The Navy would search for another supplier.[48]

The Vincent-Trammel Act included a production set-aside for the navy's aviation factory (NAF) on the assumption that the Philadelphia yard had a firm grip on its production costs. The award constituted again an undisguised bonus to civilian trade union employees.

The FDR administration added government buying power to aid and abet the nation's trade unions. If a contractor opposed dealing with a union, Swanson threatened to take its Navy work in-house. The Navy Secretary also found it useful to remind steel contractors that the Navy would restart its West Virginia armor plan if contractors flaunted the nation's labor laws. Under the guise of national security Congress employed Naval contracts to enforce its New Deal program. Senator Gerald Nye, Republican, in fact, proposed to ratchet up a corporate penalty a bit further. Nye insisted that the government nationalize the nation's shipyards, airframe suppliers, chemical firms, steel factories and commercial banks. In the 1930's, the ghost of Josephus Daniels had not been totally laid to rest.[49]

Franklin Roosevelt remained ever intrigued by the notion of a government arsenal. The President floated the concept of an Aviation Engineering Center as a benchmark to assess industry's research and development expenditures. Little wonder that the nation's aviation industry began showing indications of economic strain in the 1930's. When Congress questioned whether additional work should be assigned to the navy's aviation factory, Rear Admiral Ernest King advised the committee to move cautiously. Private contractors, he said, constituted quite a bargain. Corporate losses underwrote the navy's aviation research program.[50]

By 1937 Assistant Secretary Charles Edison detected indications that some contractors were unwilling to undertake Navy business. A financial analysis study concluded that the nation's aviation industry was "technically bankrupt," suggesting that Congress had neglected to calculate the economic cost of the administration's regulatory and tax disincentives. But all appeared well. Few lawmakers thought it inconceivable that German troops would march into Poland.[51]

Interestingly, the nation's aviation firms were not totally devoid of buyer options, and non-regulated ones at that. In the late 30's, France and Britain began purchasing U.S. planes and engines. France and England could care not a whit about the level or size of U.S. corporate earnings. The allies wanted planes yesterday. Britain, in fact, provided capital for the construction of aviation plant and equipment to the U.S. Such developments added a surreal element of to the congressional policies. Congress punished U.S. firm's capital outlays while European policies fostered U.S. investment mobilization. Exports from an unregulated overseas market nourished the research effort of domestic corporations, an early case of reverse Lend-Lease.

One result of the administration's labor support proved unsettling. Committed to collective bargaining, President Roosevelt and Congress supported labor unions when they took to the pavement to protest their corporate opponents. That was to be expected. It was disquieting, however, to witness labor immersed in its own specie of internecine warfare. As rivalry between the more conservative AF of L and the Congress of Industrial Organization, (CIO) intensified, U.S. companies found themselves caught in a labor jurisdictional crossfire.

More ominously, government wages and benefits began to exceed the earnings of private yard workers. The result was inevitable. Private yard employees went on strike in order to match the compensation of government workers. Government yards generated work stoppage at private yards. In the meantime, the President awarded contracts to build six Iowa and five Montana Class battleships at navy yards that banned adopting modern management practices. The large vessels would also require an expansion of the Panama Canal. To naval yard employees, the president's assignment constituted a vote of confidence in trade union work rules.[52]

By 1939-1940, the Navy detected an ominous trend. Private bids for Navy ship construction began to drop off by as much as 50%. Captain C.W. Fisher briefed the department's General Board on the matter. Fisher, in fact, did more. He recommended an eleven point action plan. In truncated form, Fisher's memo recommended the following:

1. The Navy Department should be permitted to assign work to private plants or government plants without reference to the present laws requiring a fifty-fifty distribution of shipbuilding as between Navy

yards and private shipbuilders. The President can invoke "National Defense" to accomplish this or it can be done by enacting Section 6 of the preliminary draft of H.R. 7665.

2. To facilitate necessary enlargements of private plants' capacities where needed, permit the Navy Department to advance not more than 30% of the contract price by enacting a clause in Section 8 of the preliminary draft of H.H. 7665.

3. In order that the shipbuilding work may be distributed to the best advantage among the various plants and sections of the country, permit the Navy Department to negotiate contracts without competitive bidding by enacting Section 9 of the preliminary draft of H.R. 7665.

4. Section 6 of the Walsh-Healey Act, approved June 30, 1936 (41 U.S.S.C., 35; Public No. 346 – 74th Congress) provided that exceptions to the prescribed minimum pay and maximum hours of labor shall be made by the Secretary of Labor upon a written finding by the head of the Contracting Department when justice or public interest, will be serve thereby. These restrictions as to wages and labor have resulted in a number of firms declining to bid on naval proposals. Up to the present, the Secretary of Labor has declined to grant exceptions when requested by the Secretary of the Navy. In order to obtain maximum competition and the best possible sources of supply, the Secretary of Labor should grant such exceptions when requested by the Secretary of the Navy, with the understanding as provided in Section 6 that any overtime will be paid for at the rate of time and a half.

5. Difficulties have arisen under the Vinson-Trammel 10% Profit Limitations Act of 27 March 1934 (48 Stat. 505), as amended by the Act of June 25, 1936 (Public Law No. 304, 74th Congress), because of the ruling by the Treasury Department regarding the acceptance of amortization charges for plant extensions and capital expenditures. Competition is being restricted because of the Treasury Department ruling that the amount to be allowed in the accepted costs for plant extensions is not determined by the Treasury Department until the termination of the contract. It is recommended that these Treasury Department Regulations be changed to provide that the cost of plant extensions necessary in connection with the performance of government contracts shall be allowed as items of cost allowable under the contracts in the amounts and under the terms as fixed upon by the Secretary of the Navy in advance of executing the contract.

6. Many clerical and technical employees (white collar employees) of the Navy Department work thirty-nine hours a week and are on a per annum salary basis and are not entitled to over-time pay. The laborers, helpers and mechanics are on a per diem basis and normally work eight hours a day and forty hours a week (five days a week). They may

not be worked in excess of eight hours a day, except in the case of an extraordinary emergency. They may be worked in excess of forty hours a week without reference to the character of the emergency. For all such overtime the per diem employees (Groups I, II and III) are allowed over-time pay at the rate of time and one half. The present laws regarding these matters are complicated and injustice results to the Group IVb employees because over-time pay cannot be granted them. It is recommended that all previous laws relating to this subject to repealed and that a new law, starting fresh, be enacted providing that the normal working hours for all civilian employees of the Navy Department be set at eight hours per day and forty hours a week, and that any time worked in excess of this be paid for at time and a half.

7. It is recommended that the monetary limitations on the employment of Group IVb employees, as now contained in the Naval Appropriation Acts, be eliminated, as these limitations complicate and hamper obtaining the necessary clerical, drafting and technical employees. All such limitations should be stricken out of the current and all future Naval Appropriation Acts and language about as follows substituted.

8. The patchwork series of laws relating to annual and sick leave for civilian employees of the Navy Department has resulted in a most complicated situation, unsatisfactory to all concerned, and in many instances unfair as regards the relative treatment of temporary and permanent employees, etc. It is recommended that all such laws be repealed and that a new simplified law be enacted to correct these difficulties. Although this may not appear to have any bearing on expediting naval shipbuilding, it does have such a bearing in that the dissatisfaction with present conditions has a definite bad effect on morale.

9. Such steps as are possible by Executive Order, moral suasion, or Act of Congress should be taken to ensure that important naval work shall be given first priority with all private manufacturers.

10. Although no legal or official Executive action is needed, shipbuilding can be expedited by making all following ships duplicates of the original design and by curtailing changes to the minimum, such as prohibiting al such changes when a vessel is 20% completed, except by special authority of the President.

11. The Bacon-Davis Act providing for prevailing wages and maximum hours of work on Federal Public Works contracts contains a provision in Section 6 thereof that this may be suspended by the President in time of national emergency. In connection with the proposed enlarged shipbuilding program, a number of Public Works items are involved, but so far as I have been able to ascertain, no particular difficulties

have arisen that would delay the construction of these Public Works on account of the requirements of the Bacon-Davis Act.[53]

In the 1930's the Navy had supported the administration's New Deal policies. By the end of the decade, some officers began to reconsider the consequences of congressional programs. Captain Fisher's memo implied that U.S. lawmakers had instituted a set of economic disincentives that punished firms, raised cost, discouraged productivity, and inhibited research in a world that was becoming increasingly troubled and disorderly. Symbolically, Captain Fisher nailed a bill of particulars on two doors, the U.S House and the U.S. Senate.

In 1940, the German army began to rearrange the geography of Europe. In the spring of the year, the French Army collapsed and Britain was forced to evacuate its expeditionary troops from Dunkirk. German U-Boats now prowled the Atlantic. Anticipating a German invasion, Britain stood alone. In June of that year, the U.S. Congress decided to increase the country's defense budget. The appropriations bill included the following clause:

> That no part of the appropriations made in this Act shall be available for the salary of pay of any officer, manager, superintendent, foreman, or other person having charge of the work of any employee of the United States Government while making or causing to be made with a stop watch or other time-measuring device a time study of any job of any such employee between the starting and completion thereof, or of the movements of any such employee while engaged upon such work; nor shall any part of the appropriations made in this Act be available to pay any premium or bonus or cash reward to any employee in addition to his regular wages, except for suggestions resulting in improvements or economy in the operation of any Government plant.[54]

Congress essentially imposed a ban on government yard efficiency. In the fall of 1940, Winston Churchill sent a technical team to the United States. As noted, the team unveiled a remarkable electronic device capable of operating at 200 megahertz/per second.[55]

Britain had constructed a home radio detection network in anticipation of a German air attack. Supported by spitfire and Hurricane fighter squadrons, that network would prevail over Goering's bombers. The home chain network validated Watson-Watt's early prediction that with radar, "Britain had become an island once more." England had given the U.S. a fifteen month advanced notice.[56]

MOBILIZATION

ANMB

In 1936, the ANMB submitted a second mobilization plan. Not unlike its earlier submission, the plan anticipated that the U.S. would create a central agency responsible for the nation's mobilization effort. Under the plan, a civilian would direct U.S. plant conversion, someone versed in management and industrial production. The central agency would balance and coordinate the needs of the military and civilian sectors. The ANMB would act as the nation's coordinator pending the creation of a central board. [57]

Congress sent the ANMB proposal to a Senate committee looking into the mobilization experience of World War I. The Nye Committee noted that chemical, steel, and banking had generated excessive profits. When members of the Nye committee read the ANMB plan, they asserted that the mobilization plan was nothing but a blue print for fascism. [58]

In 1939, the ANMB submitted still another mobilization plan to the President. The ANMB's recommendation, consistent with previous submissions, proposed a central mobilizing agency directed by a civilian administrator. In the interim, the ANMB would serve pending the appointment of a civilian director. The President rejected the 1939 plan.

FDR, on the other hand, did make two administrative moves that year. Following the German invasion of Poland, Franklin Roosevelt transferred the Army-Navy Joint Board and the Army-Navy Munitions Board to the Office of Emergency Planning, the Executive office of the president. For the first time, the ANMB had a budget. The President also created a War Resources Board to examine mobilization policy in general and the 1939 ANMB mobilization plan in particular. Chaired by Edward Stettinius, Jr., the board was composed of U.S. industrialists and college presidents.

The board affirmed the 1939 ANMB plan. Franklin Roosevelt made it clear, however, that he was less than wedded to any agency that stood between the President and the U.S. economy. Amending the 1939 ANMB plan, the board proposed several independent agencies, all of whom would report to the President. American labor leaders took issue with the board's composition and its recommendations. FDR classified the report, placed it in the White House safe, and disbanded the Board. [59]

The President proceeded to revive an agency dating back to the Wilson administration: a Council of National Defense. The council make-up included the president's cabinet—War, Navy, Interior, Agriculture, Labor and Commerce—who reported directly to the President.

The next year, 1941, the President approved the creation of the Office of Production Management (OPM). Two administrators, William Knudsen (formerly with General Motors) and Sidney Hillman, President of the Labor

Garment Workers, AFL, served as dual chairs. OPM would be advisory only. Ultimate authority rested with the White House. [60]

Congress had been relatively passive with regard to the administration's mobilization proposals. The House and Senate generally stood by as the President dealt with the 1939 ANMB plan, the War Resource Plan, the Advisory Commission to the Council of National Defense; the Office of Production Management, and later the Supply Priorities and Allocation Plan. [61]

By 1941 Congress had increased the budgets of the armed services. Concerned by the prospect of contract concentration leading to production bottlenecks, and seeking to buy time, the Army adopted a policy of negotiated contracts. The Navy's Bureau of Supplies and Accounts, on the other hand, continued competitive bidding as a matter of course.

Nor was Admiral William Leahy, Navy CNO, particularly impressed by the Army's Industrial College Program as an institution to instruct officers matters of industrial conversion, procurement and supply. Leahy insisted that no civilian was in any position to lecture a naval officer. Nor did Admiral Leahy find it useful to let educational production orders to private sector firms as an instructive, learning experience. Rather, Admiral Leahy preferred that naval officers be sent to the Naval War College where they could be trained in fleet command. [62]

Army College attendees toured industrial plants, interviewed factory managers, studied production subcontracting, reviewed labor training programs and were given a nodding introduction to financial statements, amortization and depreciation schedules. The Army solicited and welcomed Naval personnel to their Industrial War College as well.

In the meantime, Japan's treaty withdrawal in 1934 constituted an economic declaration of war. Japan wagered that its tool and die industry, its factories, its labor pool, its management, its plant, schools, colleges, and farms, could more than match the productive output of the U.S. That Japan's economy was 10% of the size of the U.S. economy apparently gave Japan's military leaders little pause. Indeed, Japan's military defined war as a clash of arms. Weapon excellence, not quantity, would trump U.S. factory output.

It was not without a touch of irony that the U.S. Congress, confronted by seismic developments on the international front, elected to pursue a program hostile to the nation's industrial sector; policies that discouraged savings, inhibited investment, depressed earnings, regulated profits, throttled innovation, penalized efficiency, punished risk taking, mandated employee costs, all the while encouraging government agencies to compete with firms in the nation's private sector.

By the end of the 1930's the U.S. was showing signs of economic stress. Labor strikes had mistatized from auto to shipbuilding to aerospace sectors of the economy. Business investment remained stagnated. In 1937, the economy experienced a recession within a depression. Congressional leaders insisted that federal labor and economic regulation was a trifling price to pay for worker democracy and industrial justice. The fact that industrial capital could elect not to commit capital funds to plant and equipment, particularly, steel, took many observers by surprise. President Roosevelt acknowledged as much when he declared that the recession of 1937 was essentially a capital strike instigated by big business.

INTER-GOVERNMENT COORDINATION

President Roosevelt elevated the stature of the Munitions Board and the Army Navy Joint Board when he ordered them to report to the executive office of the President. To that extent, Roosevelt replicated the actions of Woodrow Wilson. On the other hand, the issue of command unity remained very much an open question. In 1938, Admiral William Leahy, CNO, wrote General Craig Malin, Army Chief of Staff, that the Navy preferred a "cooperative command" arrangement as apposed to command unity. Neither the Navy nor Army, however, looked forward to taking orders of each other. Such Congressional policy was not without its unintended consequences. As Dr. William F. Atwater put it, "....In many ways it proved to be a tragic choice because it divided authority and allowed a situation to develop in Hawaii where there never existed a clear-cut line as to which service had the primarily responsibility for air defense. This muddled situation led directly to the conclusion of command responsibilities prior to and during the Pearl Harbor attack." Congress, the President and the Joint Chiefs of Staff would be required to revisit the matter of command unity, particularly in the Pacific Theater.[63]

General John A. Lejeune and President Warren Harding, Circa 1922, Gettysburg, Marine Archives.

President Warren Harding visits the carrier Langley, Circa 1923, Navy Archives.

Conversion Battle Cruiser to Aircraft Carrier, 1922, Navy Archives.

Edwin Denby and Josephus Daniels, 1921, Library of Congress Archives.

The IAM presents Daniels with an engrossed resolution of appreciation, *Machinists Monthly Journal*, 33, 5 (April 1921), 426, Georgia State University.

Great Lakes Training, left to right: Major General John Lejeune, Secretary Edwin Denby, Captain William Moffett, Circa 1921, Marine Archives.

Matrix Command

R. Adm
⇑
TYPE COMMAND

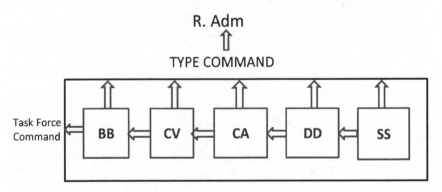

Type: Repair, Personnel, Maintenance, Overhaul,
        Administrative
Task: Tactical Training, Coordination

Matrix Command Model, Sinthy Kounlasa, UNH/WSBE, 2009.

Denby at Wireless Phone, 1922, Library of Congress Archives.

Amateur Wireless Station, Circa 1921, Library of Congress Archives.

Representative James G. Scrugham, 1935, Library of Congress Archives.

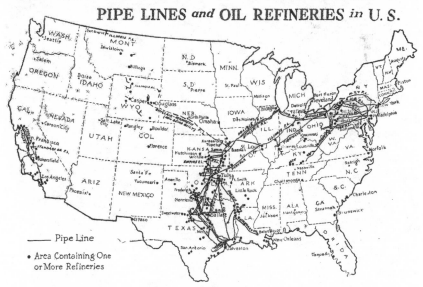

Edwin Denby papers, Burton Historical Collection, *Detroit Public Library*, Box 8, N.D.

Tank Farm, Pearl Harbor, 1941, Navy Archives.

Walter Christie and Edwin Denby, 1924, Library of Congress Archives.

Vought 024-2 Aircraft – U.S.S. *Saratoga*, Circa 1936, Navy Archives.

U.S.S. *Tennessee* supports amphibious assault, Okinawa, 1945, Marine Archives.

(III)   DEPARTMENT OF DEFENSE (WAR, NAVY).

Secretary for Defense.

Undersecretary for the Army.
Assistant Secretary.
Executive Offices:
    General Staff.
    War Boards and Commissions.
    Office of the Adjutant General.
    Office of the Inspector General.
    Office of the Judge Advocate General.
    Office of the Quartermaster General.
    Office of the Chief of Finance.
    Office of the Surgeon General.
    Office of the Chief of Ordnance.
    Office of the Chief of Chemical Warfare Service.
    Militia Bureau.
    Office of the Chief of Chaplains.
    Office of the Chief Signal Officer.
    Office of the Chief of Air Service.
    Office of the Chief of Infantry.
    Office of the Chief of Cavalry.
    Office of the Chief of Field Artillery.
    Office of the Chief of Coast Artillery.
    Office of the Chief of Engineers.
    Military Academy.
    Panama Canal.
Undersecreinry for the Navy.
Assistant Secretary.
Executive Offices:
    Office of Naval Operations.
    Navy Boards.
    Bureau of Navigation—
        Naval Academy.
    Bureau of Yards and Docks.
    Bureau of Ordnance.
    Bureau of Construction and Repair.
    Bureau of Engineering.
    Bureau of Aeronautics.
    Bureau of Supplies and Accounts.
    Bureau of Medicine and Surgery.
    Revenue Cutter Service (Coast Guard, Treasury).
    Headquarters, Marine Corps.
    Judge Advocate General—
        Solicitor.
Undersecretary for National Resources (new).
Assistant Secretary.
Executive Offices:
    Men.
    Munitions.
    Food and Clothing.
    Transportation.
    Communications.
    Fuel.
    Miscellaneous.
Joint Boards (War and Navy).
National Advisory Committee for Aeronautics (Independent).

U.S. Congress, Joint Committee, "Reorganization of the Executive Departments," February 16, 1923, 67 Cong. 4th Sess., Senate Document No. 302.

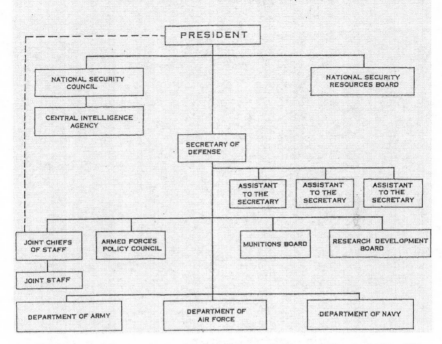

John C. Ries, *The Management of Defense: Organization and Control of the U.S. Armed Services*, (Baltimore: Johns Hopkins University Press), 1964, p. 54.

George A. Wentworth, Philips Exeter Academy Archives.

Frederick W. Taylor, Circa 1910.

Left to Right: Admiral Robert Coontz, Edwin Denby, Admiral Edward Eberle,
***Washington Post*, June 10, 1923.**

NOTES

1. U.S. Naval Administration in World War II, "The Logistics of Fleet Readiness," Vol. 20, Navy Department Library, Washington, D.C., 176. "On 3 January 1939 he (FDR) directed that no ship be sold without reporting the details to him." Also William H. Standley, *Arthur A. Agetou, Admiral Ambassador to Russia* (Chicago: Henry Regnery, 1955), 29; Paolo E. Coletta, ed., *American Secretarys of the Navy* (Annapolis: Naval Institute Press, 1980), 20.

2. Robert B. Madden, "The Bureau of Ships and it E.D. Officers," Journal of the American Society of Naval Engineers, 66 (February, 1954), 17; also U.S. Navy Bureau of Ships, an *Administrative History of the Bureau of Ships During World War II* (Washington: Department of the Navy, 1952), Vol.

3. Henry H. Adams, Witness to Power: The Life of Fleet Admiral William D. Leahy, (Annapolis: Naval Institute Press, 1985), 71-72.

4. Frank O. Brayard, *By their Works Ye Shall Know them: The Life and Ships of William Francis Gibbs, 1886-1967*, (New York: Gibbs and Cox, 1968), 108.

5. Julius Augustus Furer, *Administration of the Navy Department in World War II*, (Washington: Naval History Division, 1959).

6. NARA, RG 80, General Board Subject File, 1900-1947, GB420-2, Box 63, letter from Acting Secretary of the Navy to Chief of Naval Operations, 18 July, 1938, William D. Leahy.

7. William M. McBride, *Technological Change and the United States Navy, 1865-1945*, (Baltimore: Johns Hopkins University Press, 2001), 58.

8. John Major, "William Daniel Leahy, 3 January 1937, 1, August, 1939," in Robert William Love, ed., *The Chiefs of Naval Operations*, (Annapolis: Naval Institute Press, 1980); also Kenneth Hagan, *In Peace and War*.

9. Jan M. Van Tol, "Military Innovation and Carrier Development: An Analysis," *Joint Force Quarterly*, 77 (Autumn/Winter, 1997-1998), 104. Also "War Technology, Interwar Developments," 29, *New Encyclopedia Britannica*, (Chicago: Encyclopedia, 1998), 611.

10. McBride, 199.

11. NARA, RG 80, General Board Meetings, "Facilities for Enlarged Aviation Program, Microfilm Reel #10, 10 August, 1934, 88-137, Naval War College Library, Newport, Rhode Island.

12. L.F. Safford, "Wider Horizons" U.S. Naval Institute Proceedings, 63.11 (November, (1937) 1545.

13. A. D. Hopkins, K.J. Evans, *The First 100: Portraits of Men and Women who Shaped Las Vegas,* (Las Vegas: Huntington Press, 1999); Part I, "James Scrughman: Governor on Wheels."

14. A. Hoyt Taylor, *Radio Reminiscences: A Half Century,* (Washington: office of Naval Research, U.S. Navy, 1948) 306.

15. Taylor, 335-336.

16. Charles Susskind, "Radar as a Case Study in Simultaneous Invention," p. 238, Oskar Blumtritt Hartmut Petzold, William Asprey, Ed., *Tracking the History of Radar* (Piscataway: IEEE-Rodgers Center for the History of Electrical Engineering, 1994) According to Susskind, "...British electronics manufacturing facilities were stretched to their limits by 1940." For another view, see Correlli Barnett, *Engage the Enemy More Closely: The Royal Navy in the Second World War,* (New York: W.W. Norton, 1991) 581. "...the fundamentally deficiency lay in the limited technological capabilities of the British Radio and precision engineering industries."

17. Louis Brown, *A Radar History of World War II: Technical and Military Imperatives* (Washington: Carnegie Institution of Washington, 1999) 369; Also Jennet Conant, *Tuxedo Park* (New York: Simon and Shuster, 2002) 205.

18. Duncan S. Ballentine, *U.S. Naval Logistics in the Second World War,* (Princeton: Princeton University Press, 1947), 295.

19. Howland S. Smith, "The Development of Amphibious Tactics in the U.S. Navy," *Marine Corps Gazette,* (October, 1946), 46.

20. Timothy David Moy, "Hitting the Beaches and Bombing the Cities: Doctrine and Technology for Two Militaries, 1920-1940, Ph.D. dissertation, University of California , Berkeley, 1992.61

21. Michael Harrison, *Mulberry: The Return in Triumph,* (London: W.H. Allen, 1965), 114.

22. NARA RG 80, Annual Report of the Commander-in-Chief, U.S. Fleet, July 1938 to June 1939, Admiral C.C. Bloch, Roll #7, 25.

23. Clark G. Reynolds, *Admiral John H. Towers: The Struggle for Naval Air Supremacy* (Annapolis: Naval Institute Press, 1991), 272.

24. Jeter A. Isley and Philip A. Crowl, *The U.S. Marines and Amphibious War: Its theory, and its Practice in the Pacific,* Princeton: Princeton University Press, 1951), 36, 63.

25. Henrietta M. Larson, Evelyn H. Knowlton, Charles S. Popple, *New Horizons: 1927-1950* (New York: Harper and Row, 1971), 212; also Amphibious Training Command, U.S. Naval Administration in World War II, Commander in Chief, Atlantic Fleet, Vol. 1, 1953, VIII-18 to VIII22. Finally, Historical Collections of the Great Lakes, Bowling Green University, San Cristoban, Barnes-Duluth Shipbuilding Company, Duluth, MN 1943; HTTP://bgsu.edu-bin/xus12.c9i.

26. Thomas Wildenberg, *Gray Steel and Black Oil: Fast Tankers and Replenishment at Sea in the U.S. Navy, 1912-1992* (Annapolis: Naval Institute Press, 1996), 99-100.

27. Edgar B. Dixon, *Franklin D. Roosevelt and Foreign Affairs, March 1934 – August 1935* (Cambridge: Belknap Press, 1969), 280-281.

28. NARA RG 313, CNO, Division of Fleet Training, Esso Tankers and Fueling at Sea Gear, 28 October, 1940. Also L.S. Birnbaum, E.A. Bukzin, J.R. Saroyan, "Control of Ship Fouling in the U.S. Navy," *Naval Engineers Journal,* 79, 1 (February 1967): 78.

29. *Bureau of Ships,* Vol. III, memorandum for assistant secretary of the navy, March 16, 1938, 29.

30. Wayne G. Broehl, Jr., *Precision Valley: the Machine Tool Companies of Springfield Vermont* (Englewood Cliffs: Prentice-Hall, 1959), 161, also U.S. Congress, Senate, committee on naval affairs, Naval Expansion Program, 75th congress, 3rd sess. April 6, 1938, testimony of Admiral William Leahy Senator Bone, "...is it not a fact that the navy yards have a lot of antiquated, obsolete machinery that has been there for years, Admiral Leahy, "It is a fact that

the navy yards have antiquated and obsolescent equipment; also Bufford Rowland, William B. Boyd, *U.S. Navy Bureau of Ordnance in World War II*, (Washington: U.S. Government Printing Office, 1953), 93.

31. History Amphibious Forces, Marine Equipment Board, Meetings and Reports, Box 1, Breckinridge Library, Marine Archives, Quantico, Virginia, memorandum from 5th Marines to Commanding General, Fleet Marine Force, 26 March 1935, p. 3, Box 1.

32. Terry E. Strahan, *Andrew Jackson Higgins and the Boats that won World War II* (Baton Rouge: Louisiana State University Press, 1994), 34.

33. William Michael McBride, "The Rise and Fall of a Strategic Technology: The American Battleship from Santiago Bay to Pearl Harbor, 1898-1941," Ph.D. dissertation, Johns Hopkins University, 1989, 314.

34. U.S. Congress, Committee on the Merchant Marine and Fisheries, *The Use and Disposition of Ships and Shipyards at the End of World War II*, Graduate School of Business, Harvard University, June 1945, 160; *American Federationist*, April 1915, 292; WWW.columbia.edu/2~Rs9/BNYwelders.html, 1935.

35. NARA, RG 69, National Industrial Administration, Shipbuilding Code Hearings, July 21, 1933, Box 5250, 1015-1017.

36. Michael A. West, Laying the Foundation: The Naval Affairs Committee and the Construction of the Treaty Navy, 1926-1934," Ph.D. dissertation, Ohio State University, 1980, 334.

37. NARA, RG 69, National Industrial Administration, Shipbuilding Code Hearings, July, 21 1933, Box 5250, 1015-1017.

38. U.S. Congress, House of Representative Committee on Naval Affairs, Sunday Legislation Affecting the Naval Establishment, House bill 9653, January 27, 1936. Statement of Claude Swanson.

39. Harold G. Bowen, *Ships, Machinery and Mossbacks: The Autobiography of a Naval Engineer* (Princeton: Princeton University Press, 1954), 108.

40. Venable, 145.

41. Admiral Emory S. Land, Oral History, Columbia University Oral History Research Office, Oral History 21, Naval History Collection, 1960, 129, also Robert Earle Anderson, *The Merchant Marine and World Frontiers* (Westport: Greenwood Press, 1978), 86.

42. Gene Smith, *Rethinking the Great Depression* (Chicago: Ivan R. Dee, 1987), 127.

43. John Morton Blum, *V was for Victory: Politics and American Culture During World War II* (New York: Harcourt Brace Jovanovich, 1976), 1932; also Mary Earhart Dillon, *Wendell Wilkie, 1892-1944* (New York: J.B. Lippincott, 1952), 49-50.

44. Ellis W. Hawley, *The New Deal and the Problem of Monopoly* (Princeton: Princeton University Press, 1966).

45. Jacob A. Vander Muelen, *The Politics of Aircraft: Building an American Military Industry* (Lawrence: University Press of Kansas, 1991), 6-7.

46. U.S. Congress, Senate, Special Subcommittee Investigating the Munitions Industry, *Munitions Industry*, 64th Congress, 2nd Sess. February 6, 1936, 95590.

47. Muelen, 129.

48. Curtis Alan Utz, "Carrier Aviation Policy and Procurement in the U.S. Navy, 1936-1940," Master's thesis, University of Maryland, 1984, 19.

49. Ferrell, 210.

50. Irvin Briton Holley, Jr., *Buying Aircraft: Material Procurement for the Army Air Forces* (Washington: Department of the Army, 1964), 277.

51. Jacob A. Vander Meylen, *The Politics of Aircraft: Building an American Military Industry* (Lawrence: University Press of Kansas, 1991), 192.

52. McBride, 179.

53. NARA RG 80, General Board File, subject file 6B 420-2, memorandum from Director of Shore Establishment (C.W. Fisher) to General Board, 10 May, 1040, Box 63.

54. United States Statutes at Large, 1939-1941, (Washington, D.C.: United States Printing Office, 1941): 39.

55. Corelli Barnet, *Engage the Enemy More Closely: The Royal Navy in the Second World War* (New York: Norton, 1991), 58.

56. Robert Buderi, *The Invention that Changed the World* (New York: Simon and Shuster, 1996), 58.

57. Albert A. Blum, "Birth and Death of the M-Day Plan," Harold Stein ed., *American Civil-Military Decisions* (Tuscaloosa: University of Alabama Press, 1963), 68, "Union Leaders remained prominent among the critics of the Imp's during all of the 1930's," p. 69.

58. Blum, 68.

59. Jordan A. Schwarz, *The Speculator: Bernard M. Baruch in Washington, 1917-1965* (Chapel Hill: University of North Carolina Press, 1981), 369.

60. Schwarz, 375-376.

61. Henry A. Adams, *Witness to Power, the Life of Fleet Admiral William D. Leahy* (Annapolis: Naval Institute Press, 1985), 78.

62. Military History, 1977, 463; also William Felix Atwater, United States Army and Navy Department of Joint Landing Operations, 1898-1942," Ph.D. dissertation, Duke University, 1986, 113.

63. Marvin A. Kreidberg, Merton G. Henry, "Industrial Mobilization Planning Between the Wars," Charles R. Shrader, ed., *United States Army Logistics, 1775-1992, An Anthology* (Washington: Center for Military History, 1977), 463; also William Felix Atwater, United States Army and Navy Department of Joint Landing Operations, 1898-1942," Ph.D. dissertation, Duke University, 1986, 113.

Chapter Five

The Pacific Campaign, 1941-1945

In November 1943 the U.S. launched a central-Pacific campaign that took the fleet 5000 miles to the Gilberts, the Marshalls, the Carolines, Peleliu in less than twelve months. The central Pacific offensive provides a focal point to revisit the administrative decisions of the Harding administration.

ADMINISTRATIVE OVERSIGHT

Franklin D. Roosevelt transformed the Joint Army Navy Board into the Joint Chiefs of Staff (JCS), whose membership included General Henry Arnold, Aviation; General George Marshall, Army; Admiral Ernest King, Navy, and Admiral William Leahy, chairman and the president's representative. Hammering out U.S. strategic planning, the Joint Chiefs agreed to send only unanimous recommendations to the President.

President Roosevelt included Henry Stimson or Frank Knox, his Secretary's of War and Navy, from participating in matters of military planning. Nor did Admiral King find it convenient to brief James Forrestal, Under Secretary of the Navy, on matters of strategy and tactics. Occasional lunches between John McCloy, Assistant Secretary of War, and Forrestal kept the latter somewhat up to date.

During the war the Navy's chain of command occasionally ran at cross-purposes. As the Fleet's Commander-in-Chief, Admiral Ernest King reported to the President; as Chief of Naval Operations, Admiral King reported to the Secretary of the Navy. Although Secretary Knox had recommended that Admiral Ernest King be promoted to Chief of Naval Operations, their relationship became strained overtime. Ever suspicious of civilians, King avoided contact with both the under and Assistant Navy Secretaries.

The material bureaus, independent of direct CNO oversight, continued to engage in their respective buying practices. A month after the Japanese attack on Hawaii, Under Secretary James Forrestal set up the Office of Procurement and Materials (OPM) as a way to synchronize the bureau's purchasing activities. Forrestal later appointed Admiral Samuel Robinson to direct OPM's bureau oversight. On more than one occasion, Admiral King attempted to pull OPM within his management wing. The President, however, blocked King's moves.[1]

Prior to December 7, 1941, Admiral Harold Stark served as Chief of Naval Operations; Admiral Ernest King commanded the U.S. Atlantic Fleet; Admiral Husband Kimmel, the U.S. Pacific Fleet. Following the Japanese Hawaiian strike, the President relieved Admiral Harold Stark and sent him to London to head U.S Navy Operations in Europe. Roosevelt appointed Admiral Chester Nimitz as Kimmel's replacement in the Pacific. By March, 1942 Admiral Ernest King had consolidated his administrative power as Chief of Navy Operations (CNO) and Commander in Chief of the United States Fleet. (CINCUS), altered as Cominch. At the same time, Admiral King brought operation's Division of Fleet Training Division under the Commander-in-Chief's office.[2]

According to the historian Eric Larrabee, Admiral King viewed the navy's shore establishment as "...too complex for the creaky pre-war machinery of the bureaus to cope with...," King wanted direct authority over the Navy's Bureau Chiefs. The President felt that such a move required congressional legislation and time was in short supply. Instead, the President reassured King that he could replace any Bureau Chief whose performance he found unsatisfactory.[3]

As Cominch, Admiral Ernest King exercised coordinating authority over the operating fleet the bureaus and shore establishment, a remarkable expansion of naval responsibility. Actually, the consolidations process had commenced in 1940. Rear Admiral Julius A. Furer, observed that, "In addition to the duties now prescribed by law the Chief of Naval Operations shall, under the direction of the Secretary of the Navy, be charged with the coordination of the functions of the naval establishment afloat, together with the determination of priorities relating to repair and overhaul of ships in commission or about to be commissioned." As Furer put it, "No law was more important to the efficient administration of the navy department in World War II than the act of June 20, 1940."[4]

In the early months of the war, fleet logistics fell under the jurisdiction of the Deputy Chief of Naval Operations. When Admiral King did release a specific military timetable, the lead time required for machine retool set up, production schedules and ship conversion, required private contractors to embark on a crash program. Such tardiness prompted Kent Algier to observe that "...even under wartime conditions the direction of the logistics plans division in CNO was not given access to the "top-secret dispatch board of the Commander-in-Chief until late in 1943, two years after the war began." Private contractors had to overcome the navy's tendency to inflict secrecy upon itself. It would take time to recognize that fleet tactics could not be entirely divested from the nation's supply chain.[5]

The urgency and complexity of naval engagements coerced the department to adopt modern business practices. The Bureau of Supplies and Accounts and the Bureau of Ships, for example, was unable to estimate how long a war ship could operate before requiring additional fuel and food. Suspecting organizational lacunae, Secretary Frank Knox enlisted the advisory services of Booz Allen Hamilton. A Chicago firm providing scientific management advice, Booz at first straightened out the navy's telephone system. The firm next addressed the interface between war planning and fleet supply. On paper, the War Plans division routed its material requirements to the department's Bureau system. In practice, Booz concluded that the liaison between Commander-In-Chief and the Office of Operations was at best episodic. Booz Allen recommended that Bureau personnel be assigned to the Operation's office. Subsequently, "...representatives of the Bureau of Supplies and Accounts, Yard and Dock, Ordnance, and Ships were ordered to duty in the plans Division of Operations." Coordination between Operations and the Bureaus was thus achieved by personnel assignment.[6]

Harding/Denby

In the early 1920's, Edwin Denby had attempted to broaden the coordinating authority of CNO's Operations. General order 433 accorded the CNO cognizance over ship repair and alteration, authority intended to ensure that the Navy's Bureau system responded to the needs of the fleet. In the late 20's and early 30's, General Order 433 fell into disuse. Nor was President Roosevelt particularly helpful. By siding Navy's Bureau Navy's Bureau Chiefs the President weakened Operation's the Navy's Bureau Chiefs the President effectively weakened bottom-heavy.[7]

BALANCED FLEET

Pacific War

Japan's Pearl Harbor attack forestalled the possibility of an early U.S. battleship response. The Navy first concentrated on resuscitating its damaged warships. In the interim, Nimitz sent carriers to the central Pacific on sporadic raids. Nevertheless, a General Board Advisory letter of February 1942 emphasized the preeminence of the battleship. Commenting on the loss of a British battleship and cruiser to Japanese land aircraft in the South China

Sea, the Board assigned the sinking to poor seamanship. The Board did acknowledge that in the future battleships did require air protection. At the same time, the Board recommended that the government activate the Navy Department's armor plant in West Virginia.[8]

In the early months of 1942, German U Boats enjoyed a field day sinking American oil tankers plying their trade along the Caribbean and the U.S. Atlantic coast. Henry L. Stimson, Secretary of War, suggested that the Army employ its B-24 bombers as an anti-submarine weapon. The Navy proceeded to ramp up destroyer and destroyer escort production, supplemented by Navy B-24's. The deadly effectiveness of German U Boats during World War I had somehow been neglected or forgotten.

U.S. carriers, delivering aircraft to Pacific bases, were absent from Hawaii on Sunday morning, December 7th. In early Pacific war skirmishes, the Lexington Class carriers still made their mark. Stephen Roskill, the British historian, observed that the "...U.S. Navy was extremely far sighted or perhaps extremely fortunate in standing for the conversion of the Lexington and Saratoga. For those ships exercised a decisive influence on the recovery of allied sea power in the Pacific in 1942." As a mobile strike force, the carriers were able to purchase time while U.S. forces waited for military assets to roll out of U.S. factories and yards.[9]

The naval historian, Mark Campbell, added that "...it was the experience of the Lexington and Saratoga in the decade before the Second World War that laid the foundation of the great task force;" Campbell noted that the Lexington Class carriers would not have been built had it not been for the 1922 Washington Naval Agreement. Later, Admiral King observed that had the Lexington Class remained as battle cruisers, they would have been obsolete by 1940.[10]

Admiral King did place a hold on 45,000 ton battle cruiser construction. The navy now sought aircraft carriers. A shortage of steel posed as a key limiting factor. Heavy warship construction was not without its opportunity cost—transports, cargo vessels, escort carriers, fleet auxiliaries—not built.[11]

Skeptical of a matrix command structure, Admiral William Pratt reoriented the fleet back toward a command type protocol in the early 1930's. Ten years later, Admiral James O. Richardson, Commander-in-Chief, U.S. Fleet, reintroduced a task force command system. In Admiral Richardson's words;

> In 1940, the need was very greater for inter-type training to give subordinate flag officers in the fleet operational functions and responsibilities fully capable to the probable operations of war. This requirement coupled with the further need to train units of ships and aircraft for carrying out specific parts of the War Plans, led to the establishment in June 1940 of numbered task forces, the early forerunners of the actual task forces and task fleets of World War II.[12]

By 1943, the Essex class carriers began joining the Pacific fleet. The Battle Force had transitioned to a Fast Carrier Task Force, a matrix command structure subdivided into smaller task groups. Professor Robert Albion, the naval historian, observed later that the task force command structure contributed to the Navy's administrative success in the Pacific. [13]

After a lapse in the late twenties, the Navy revived its circular cruising disposition. Carriers resided at ring's center—battleships, cruisers, and destroyers, submarines—stationed in outer concentric rings. In ring formation, the entire fleet turned into wind for carrier launching and landing requirements. In addition to adopting the U.S. circular formation, the British fleet employed another U.S. innovation by designating an air officer, "...in tactical demand (OTC) by day, this responsibility reverting to the overall commander in a battleship or cruiser by night." [14]

The Coral Sea and Midway engagements (1942) underscored the ascendance of carrier aviation. In both battles apposing warships never saw each other. Carrier production now became matter of urgency. The Navy expedited Essex carrier construction, converted nine Cleveland class cruisers into light carriers, and recycled four Standard oil tankers into escort carriers.

Escort carrier construction commenced in the summer of 1942, although that decision was by no means assured. Originally Admiral King and his colleagues vetoed Henry Kaiser's proposal to convert merchant hulls into small carriers (CVE's). President Roosevelt reversed their decision. Designed to meet the threat of enemy submarines, the CVE's would be employed as air support in amphibious operations and serve as a screen for fleet oilers. Though Admiral Leahy remained singularly unimpressed with destroyer escorts, the U.S. later employed destroyer escorts as an effective anti-U-Boart weapon.

In the early days of the Pacific war, Japan's Naval aviation proved formidable. Japanese pilots had honed their air tactics during their China incursion. The Mitsubishi fighter, the Zero, proved a remarkable machine, more agile than comparable U.S. planes. U.S. pilots had to compensate for the Zero's speed, maneuverability and climbing ability. With the arrival of the Grumman Hellcat, the U.S. Navy possessed a fighter that more than matched its Japanese counterpart. And though somewhat dated, the navy's Dauntless diver-bomber provided a margin of victory in the Coral Sea and Midway engagements. U.S. Devastator torpedo pilots, on the other hand, sacrificed themselves at the battle of Midway, prompting one historian to argue that the real culprit was not the Japanese Zero fighter but rather congressional procurement regulations. [15]

The Bureau of Aeronautics (BUAER) by now had consolidated aviation decision-making within one organization. One British historian concluded that BUAER had become the prime mover in air tactical innovation between the two world wars. Another commentator asserted that U.S. Naval aviation

"...revolutionized U.S. naval strategy and tactics." As to the appointment of Rear Admiral William Moffett as BUAER Chief, Steven Roskill concluded that "No other single step was to prove as fruitful for the United State Navy and have such far-reaching effects, as this appointment." BUAER thus promoted specialized aircraft designed to withstand the punishing environment of heaving carrier decks. Powered by air cooled radial engines, fueled by high octane gas, endowed with armament protection, subsequent generations of U.S. planes would overwhelm Japan's carrier aircraft.[16]

The Royal Navy, by contrast, attempted to recycle its magnificent land aircraft, the Spitfire and Hurricane, into carrier planes. The transition proved an unhappy experience. RAF planes suffered damaged undercarriages, restricted operating range and limited storage space due to their fixed wings. By the end of the war America would supply the bulk of England's naval aircraft together with some forty U.S. escort carriers.

In electing to pursue their own power plant, the Army favored liquid cooled engines, the Navy, the air cooled radial. On its face, engine procurement appeared to be an exercise in service duplication. Professor Robert Schlaiffer, however, suggested that "the existence of two individual agencies meant that the mistakes of one were corrected in a surprisingly large number of instances by the actions of the other." Enjoining an aviation merger permitted each service to tailor aircraft tactics to a specific tactical requirement.[17]

Flight training programs swelled the ranks of Naval pilots during the war. By the end of the Pacific campaign, naval aviation had undergone a personnel transformation. Virtually all pilots were reserve rather than regular officers. That the Navy was largely a reserve organization was occasioned when Chester Nimitz ruled that regular and reserve officers should wear identical uniforms.[18]

The Navy had to confront the technical convergence of electronics and communications. Proximity fuses, radar fire control, radar plots and Combat Information Centers (CIC) played a critical role in subsequent Pacific encounters—Coral Sea, Midway, the Philippines Sea, Leyte Gulf, Okinawa. In the latter months of the war U.S. destroyers employed a radar/communication defense system against Japanese Kamikaze's attacks. Tactical command officers tended to move to the ship's CIC center, operated by fighter direction officers (FDO) who, in civilian life had been teachers, advertisers and salesmen.[19]

U.S. submarines came into their own during the Pacific war, although that development did not occur overnight. The production of 200 submarines accompanied by new tactics isolated Japan from its source of food, oil and raw materials. Operating in small groups, U.S. submarine destroyed upwards of 50% of Japan's merchant fleet, a process delayed by faulty torpedoes.[20]

Harding/Denby

The Washington treaty system rescinded President Wilson's 1916 naval construction program and brokered a U.S.-Japanese naval truce that lasted some fifteen years. To this day, the treaty system remains controversial. One naval historian called the treaty diplomatic folly. Another historian referred to the treaty as a defensive crime. Mark Campbell, on the other hand, concluded that, "the essential foundation for American naval success in World War II was laid by the various naval treaties in effect from 1922 to 1936 by their prohibition of battleship development and encouragement of carrier construction."[21]

It is sometimes useful to consider the decisions not taken. Warren Harding did not consolidate Army and Navy air into an independent aviation unit. Britain's RAF experience suggests that had the Army Air Corps taken over naval aviation, U.S. carrier aviation might well have atrophied. Dr. Steven Rosen, in fact, observed;

> The popular account of Billy Mitchell and the navy overlooks the fact that Mitchell fought but failed to prevent the establishment by President Harding of the navy's bureau of aeronautics, an event that all historians point as crucial to the successful development of carrier aviation.[22]

Dr. David McGregor added that "every student of naval aviation from Roskill to Marden, Till to Hone and Mandeles, has recognized the advantage of the U.S. Navy over the Royal Navy in possessing an independent air arm."[23]

In emphasizing its strategic bomber, Britain's Royal Air Force siphoned resources away from the Royal Naval's fleet. The RAF controlled naval aircraft purchases, dispensing RAF hand-me-downs to the operating fleet. Committed to a horizontal bombing protocol, the RAF was somewhat tardy in developing dive bombing techniques.

The Harding administration attempted to simplify to the maze of U.S. government procurement rules. More critically, the President Harding had assigned commercial aviation development to Herbert Hoover, Department of Commerce. As cabinet officer and later President, Herbert Hoover would serve as nation's aviation's benefactor.[24]

Harding's aviation policy was driven by his conviction that market forces would energize engine and aircraft design. That wager would bear fruition in the early 1930's. By that time, civilian planes surpassed military planes in speed, range, and carrying capacity, an innovation cycle that, after 1937, picked up speed. In four short years aviation design life had collapsed from years into months. Naval surface officers, by contrast, tended to calculate dreadnaught life by the decade.[25]

Interestingly enough, the Washington Naval Treaty resurfaced during the presidential election of 1944. In October of that year, Franklin Roosevelt asserted that the Harding administration had "scuttled" the fleet, apparently referring to Harding's cancellation of ten battleships and six battle cruisers in 1922. That FDR remained wedded to the battleship can be seen by his push to build six Iowa class and five Montana Class battleships together with six Alaska Class battle cruisers. It was almost as if the President was driven to resurrect Wilson's dreadnaught program of 1916. The Navy cancelled five Montana Class ships, and Admiral King halted the construction of the Alaska class battle cruisers. Nevertheless, Dr. William M. McBride concluded that in the decade of the 30's, President Roosevelt "...had rebuilt the American navy around the battleship strategy." That indictment may have been overly harsh. Congress was equally complicit in funding the Navy's battleship program. [26]

AMPHIBIOUS OPERATIONS

Pacific War

Between the two world wars, the Marine Corps Expeditionary Force experienced several changes. By November 1943, the Corps encompassed aviation, landing craft, amphibious ships, battleships, cruisers, destroyers, submarines, cargo, stores ships, field artillery. The Gilbert amphibious assault of November 1943 set a generic pattern, though the learning process was not inconsequential. Once a U.S. fleet task force had isolated and controlled an island's air and sea space, the fleet assumed a protective stance against enemy interdiction. Heavy gunship fire and carrier aviation softened an island's defenses. Landing craft, transports, and cargo vessels, LST's (Landing Ship Tanks) unloaded LCM's (Landing Craft Medium); and later LSD's (Landing Ship Dock) disgorged Marine and Army units. The Navy's submarine force intercepted enemy transports bearing supplies and reinforcements.

In June 1944, Admiral Raymond Spruance's Fifth Fleet confronted a tactical dilemma; seek out the enemy fleet or assume a posture of defense. To the frustration of his carrier task force commander, Spruance elected the latter, accepting the fact that amphibious operations constituted his overriding priority. During the battle of the Philippine Sea, U.S. forces destroyed some 300 aircraft aided by radar directed fire and time delayed radio fuses. Reflecting on lessons acquired from interwar fleet exercises, Admiral Spruance recalled, "...we found out that there had been a tendency to decide what an enemy was going to do and lose sight of what he could do. I have seen this happen in fleet problems at sea and it is very dangerous." [27]

In 1944, Army forces landed on the Philippines Island. Given a somewhat ambivalent mission statement, Admiral William Halsey spotted Japanese carriers northeast of the invasion beach, vacated the landing area, and pursued what turned out to be a decoy of Japanese carriers-devoid of aircraft.

With the invasion beach exposed, a Japanese fleet suddenly reemerged from San Bernadino straits.[28]

Leyte Gulf would emerge as the war's largest Pacific naval engagement. The sacrifice of U.S. Navy destroyers, escort carriers, uncommon personal courage, coupled with Japanese officer fatigue, saved the day for U.S. forces. Admiral Howard Vickery, U.S. Maritime Commission, later wrote a letter to Henry Kaiser, stating, "We owe this day to you and your carriers."[29]

During the Army Philippines campaign, Marine ground observers directed tactical air support. Army ground troops solicited and welcomed Marine air's timely response.[30] Air Force doctrine, by contrast, contributed to in an unfortunate event in Sicily when their planes unilaterally adopted radios incompatible with infantry ground sets. So infrequent were U.S. Army aircraft that when transports flew over Sicily's Gela beach bearing paratroopers, U.S. artillery shot down Army planes, thinking they were enemy aircraft. It was almost as if the U.S. Army Air Corps had over-subscribed to Sir Hugh Trenchard's commitment to the strategic bomber.[31]

The U.S. Navy did take advantage of a British innovation in the Pacific War, an amphibious command ship (AGC). Embedded with teletype terminals, telephones, telegraph, sonar, radio and radar, the AGC served as a nerve center that directed artillery, landing craft, supply vessels, landing craft and beach parties. The AGC became an electronics information system afloat, a coordinating node essential to any island assault. As one report put it, "They [AGC's] are able to turn air and gunfire support on and off like a hose."[32]

At Tarawa, the Marines encountered static problems with AM radios. Two months later, the Marines operated FM sets. Amphibious trucks and tracked landing craft delivered ordnance and supplies to front line forces—a logistics flow that reduced traffic tie up and beach congestion.[33]

Throughout the Pacific war, the Marines requested their own carriers and planes. They would get them in 1945. The coordination of Marine air support prompted Professor Gerald F. Kennedy to conclude that the Aviation Ground Officer's School and the Air Infantry School constituted "...the most important schools at Quantico."[34]

By the last year of the war, Marine-tank teams had congealed into a tactical partnership—radio serving as the coordinating interface. The Japanese commander at Okinawa acknowledged as much when in his last dispatch he wrote, "the enemy's power lies in its tanks."[35]

Within less than a year U.S. Pacific forces had leaped-frogged the far reaches of the central and western Pacific. By August, 1945, some 40% of the fleet was assigned to amphibious operations employing weapons built *after* rather than before Pearl Harbor. Navy pilots were largely reservists and some ninety five percent of the fleet's auxiliary ships were also commanded by reservists.[36]

Prior to the war, Japan may have viewed the Pacific as a moat insulating its home islands from the reach of enemy forces. U.S. amphibious forces shattered that perception. After the war, J.F.C. Fuller, the British historian, noted that the Pacific ocean that had become Japan's "deadly enemy," adding that amphibious warfare "...was the most far reaching tactical innovation of the war."[37]

After the war, fleet Admiral Chester Nimitz succeeded Admiral King as Chief of Operations. As CNO, Nimitz redesignated the mission of Navy's battleships. They were assigned as fire support for amphibious operations, suggesting that John A. Lejeune's 1923 Naval War College lecture had indeed been a prophetic one.[38]

Historians generally agree that the U.S. Army, Navy, Marines, and Japan, all made contributions to amphibious warfare. It was the U.S. Marine Corps, however, that codified joint operation's precepts that, in turn, would redound to the benefit of the Army. The Army's First, Third, Seventh and Ninth Army Divisions participated in amphibious operations in Oran, Casablanca, Sicily, Anzio, Normandy, Kwajalein, Leyte and Okinawa.

Harding and Denby

President Harding and Edwin Denby accepted John Lejeune's vision of the U.S. Marine Corps. A Marine reservist, Edwin Denby's appointment of Lejeune as commandant proved to be an inspired choice. Lejeune, in turn, placed Lieutenant Colonel Earle Ellis in charge of Marine planning operations. Long intrigued by Japan's ambitions in the Far East, Ellis produced a U.S. offensive plan premised on island hopping across the central and western Pacific.

Prior to the first world war, the Marines experimented with an Advance Base Force. After the 1918 Armistice, Secretary of the Navy, Josephus Daniels approved the Marine's formation of an east and west coast Expeditionary Force (MCEF). It was the Washington naval treaty, however, that inspired Lejeune to define the Marine Corps as a mobile force dedicated to securing Pacific bases for the U.S. fleet.

As Secretary, Edwin Denby supported Lejeune's amphibious program. Lejeune envisioned the MCEF as a self-supporting enterprise, melding ground troops, artillery units, landing craft, battleships, naval and marine aviation, radio signal units, a tank battalion and supply logistics into a formidable weapons system. Major General John Russell's "Fleet Marine Force" sealed a CINCUS administrative bond with the fleet in the 1930's.

Denby approved of the formation of a Control Force designed to assist the Marines and Army in any joint operation. Composed of cruisers, destroyers, transports, cargo vessels, lighters, and submarines, the Control Force replicated the matrix command structure of the Battle Fleet.

The Caribbean exercises of 1923-24 heralded a series of lessons learned, assets required, tactics revised, a timeless exercise in trial and error. By the mid 1930's, annual fleet exercises (Flex's) continued as a learning platform not only for the Marines but for the Fleet as well. Digesting the experience, the Marines sharpened their tactical agenda—battleship gun support, landing craft, amphibious tanks, close air support, combat loaded, transports, a logistics quarter-master specialized attack planes, shore parties, an amphibious command ship. Wireless (radio) was now accepted as essential to any sea, air and ground operation.

LOGISTICS

Pacific War

However successful in immobilizing the U.S. battle line, Japan's Pearl Harbor attack neglected to destroy the island's infrastructure of oil tanks, drydocks, repair facilities, machine shops, storage depots. Admiral Husband Kimmel speculated that had the Pearl Harbor oil tank farm been destroyed, the Pacific fleet would have had to relocate to the U.S. west coast, some 2200 miles east. Admiral Nimitz estimated that the destruction of oil tanks would have prolonged the war by an additional two years. [39]

In May, 1942, Pearl Harbor's capital infrastructure now paid off. Damaged at Coral sea, the *Yorktown* limped into Pearl Harbor. Within three days, machine shops, repair units and personnel patched up the carrier in time to make a telling contribution to the Navy's Midway victory. Midway marked the high point of Japan's offensive drive. It was at Midway that Japan lost its wager that the war would be a short, quick, easy enterprise. On the contrary, the Pacific war would be extended and prolonged, underwritten if not determined by American industrial power. U.S. Naval strategy had now become a dependent variable.

By the fall of 1943, Pearl Harbor emerged as the operational center for fleet aviation, submarine force, service squadron, Marine Corps. A logistics' node, Hawaii nourished the navy's offensive campaign in the central Pacific. After the war, historians would describe Pearl Harbor's oil storage facility as an act of foresight of "incalculable importance." William Halsey's hit and run tactics obviously depended on Pearl Harbor oil. [40]

During the U.S. Central Pacific offensive, commercial tankers shuttled between the Caribbean east coast, west coast and Hawaii. Some 300 commercial tankers would transit 5,000 miles to deliver oil to Pacific lagoons and anchorages, a communication line that as a logistic target exposed the tankers. Indeed, so many merchant tankers delivered oil that the Navy set a 40 mile gap between east-west traffic to avoid tanker collisions.

Spotted by German submarine officers, Germany loaned Japan a submarine as an instructive weapon of choice. Japan, however, much preferred to direct its submarines against U.S. warships. Japan also used them as freighters, ferrying supplies to troops in Japan's scattered island outposts. In so doing, Japan neglected an exposed navy supply train. German analysts validated Admiral John K. Robison's earlier concern that Japan might and could isolate Pearl Harbor from the U.S. mainland.[41]

The Navy's Construction Battalions, a division of the Bureau of Yards and Docks, landed with Marine, Navy and Army contingents. LST's delivered tractors, bulldozers, graders, rollers, construction material permitting the CB's to build airfields, roads, wharfs, buildings, storage facilities. Their work transformed an enemy island into a launching pad for a subsequent island attack. By 1945 250,000 C.B.'s representing 60 different trades had built some 600 advanced bases during the Pacific War. On Guam alone, the CB's constructed 93 miles of roads.[42]

After three days of fighting, U.S. surface vessels returned to Hawaii for replenishment and stores—a route that took them out of action for as long as 10 to 12 days. Viewing the round trip an exorbitant waste of time, Admiral Raymond Spruance ordered the creation of mobile base force to supply the Fifth Pacific Fleet on the move. Employing oilers, ammunition ships, fleet tugs, Sevron #6 rendezvoused with the fleet in designated forward areas. Screened by destroyers and escort carriers, the service squadron #6 was designated task group 30.8 as part of a matrix command structure.[43]

The U.S. Navy eventually turned their fleet oilers into multi product service suppliers. While fueling at sea, naval research technicians transited from oiler to destroyers, cruisers, battleships, carriers, checking, repairing, and instructing ship forces on the latest radar equipment update. The fleet had come a long way from a fixed base dependency.[44]

As a floating naval base Sevron #10 patched up damaged vessels and made them whole again, an assignment dependent upon individuals who could diagnose, improvise, and repair ships absent design plans or drawings. In some instances repair units fixed warships faster than their continental counterpart. Not incidentally, mobile dry-docks liberated the fleet from stateside civilian yard work rules. Mobile bases were especially critical in repairing ships damaged by Kamikaze attacks. Captain W.D. Puleston later observed;

> Off Okinawa in the spring of 1945, approximately 250 ships were hit by Kamikaze. Many of the smaller ones sank; many of the larger were severely damaged. And except for the floating bases and the steady stream of replacements the war would have been prolonged indefinitely.[45]

Over time any coordination between carriers and mobile supply squadrons mandated close staff planning. The Pacific war eroded administrative lines and boundaries as carrier task forces and logistic supply staffs blended into a virtual planning organization. By extending the forward reach of U.S. operating forces by some 6,000 miles mobile logistics compressed the navy's offensive time table, a cycle that disrupted Japanese defense plans.[46]

The volume and quantity of petroleum delivered to the Pacific war accounted for over half of the navy's annual tonnage usage. Oil tank farms on both coasts served a dual war front. In the Pacific, Pearl Harbor emerged as a surge tank for central Pacific operations.[47]

The nation's domestic pipelines stood as a dual use infrastructure. During the war, U.S. built two pipelines that stretched from the southwest to refineries in the midwest and northeast, supplementing commercial petroleum delivery systems including Sinclair's 900 mile feeder line from Wyoming. Although the Teapot Dome reserve had long been shut down, an adjacent reserve, Salt Creek, remained productive as an oil source throughout the 1940's.

As an addition to oil transportation, the big inch pipelines released 70 merchant tankers. Two dozen tankers joined the Pearl Harbor's shuttle, delivering oil to a floating inventory base at Sevron #10 anchorages. Neglected by Congress in the 1920's, the Navy perceived its oilers and merchant tankers as indispensable logistic assets.[48]

The number of ships assigned to the Pacific Service Force grew exponentially during the war. In 1940, the Base Force averaged some 50 vessels. By 1945 the Service Force Pacific included to 2930 ships, including 300 oilers converted from commercial tankers.[49]

U.S. combat experience erased clean demarcations between auxiliary and combatant vessels as well. In 1941, the Navy labeled a transport vessel as an auxiliary ship, a year that General H.H. Smith insisted that transports employed in amphibious operation no longer deserved that designation. By 1945, cargo and transports were termed "attack" ships and earned battled stars. Despite their official designation, mobile dry dock and repair operations took on a quasi-combatant status as well. Some Naval reservists, in fact, defined them as offensive assets.[50]

The status of the nation's petroleum industry shifted during the war. After the U.S. asked Standard Oil of California to manage naval petroleum reserve number one, the company adopted Albert Fall's unitized oil drilling protocol as its mode of operation. Following the Japanese attack on Hawaii, the Navy secured the services of Standard Oil's fleet manager to straighten out Pearl Harbor's traffic problems. Harold Ickes, Interior Secretary, asked Ralph Davies, a vice president of Standard Oil of California, to coordinate the nation's petroleum production, refining transportation and distribution activities. Standard Oil reimbursed the pay disparity between Davies' government sala-

ry and his corporate income. In short, a private corporation paid a government employee's salary. The U.S. Senate did not order President Roosevelt
to relieve Ralph Davies as Assistant Administrator of Nation's War Petroleum Agency. And among other activities, the nation's oil companies diversified into high octane fuel and synthetic rubber production.[51]

Teapot Dome continued to cast its shadow over U.S. oil policy. In the
1920's, Edward Doheny's Pan American Petroleum Company sold its Venezuelan operations to Indiana Standard. Later, New Jersey Standard Oil acquired Indiana's oil Venezuelan assets, an Aruba refinery and access to a
Maracaibo oil fleet. In the late 30's Britain purchased three Maracaibo tankers, converting them into amphibious landing ships, (the Winnie) in time for
the 1942 North Africa Campaign. Redesigned by the U.S. Bureau of Ships,
the Maracaibo shuttle tankers became a distant relative to the LST (Landing
Ship Tank).[52]

The Navy asked the Barnes-Duluth Ship Building Company to construct
shallow drafted tankers destined for the Maracaibo oil fields. They would
join Manitowoc Wisconsin built submarines in traversing the Mississippi
River to the Gulf of Mexico on route to Europe or the Far East.[53]

The Navy made singular progress in its search for an effective antifouling
paint. By 1940, the Navy found that hot and cold plastic paint doubled the
time a warship could operate before being compelled to return to a Navy
Yard dry dock. Originally, the antifouling paint was reserved for U.S. warships. Over time, fleet auxiliaries would qualify for hull coating as well. By
the spring of 1945, some thirty five merchant tankers received hot plastic
paint on their hull bottoms. Japanese naval officers assumed that U.S. warships could operate no longer than nine months before withdrawing to a navy
yard. On that basis Japan doubted that the U.S. could retain a concentrated
fleet in the western Pacific. According to Dr. John R. Saroyan, "it was a real
surprise to the enemy when aircraft carriers, battle wagons and heavy cruisers that were supposed to be in dry-dock started pouring out armor-piercing
shells…" Saroyan added that U.S. antifouling paints had shortened the war
by 18 months.[54]

Harding/Denby

Denby signed off on Daniels' Pearl Harbor defense plan that, among other
requirements, included a fuel storage depot, an ammunition depot, a machine
shop, channel dredging, a naval air station, a submarine base. To sustain the
yard's civilian work force, Admiral Coontz ordered the Pacific Fleet to take
its repair and overhaul work to Hawaii. After resigning as secretary, Edwin
Denby asserted that Pearl Harbor's oil tank farm would be available to the
fleet when and if necessary. Pearl Harbor oil tanks fueled the navy's 1942 hit

and run tactics in the central Pacific. If nothing else, Denby's oil tank farm had been a farsighted investment in spite of the U.S. Supreme Court.

Although the Navy did not formally activate a Construction Battalion in the 1920's, the Bureau of Yard and Docks at least conceptualized its mission as an adjunct to Operation's Orange offensive plan. On the other hand, the fleet did anticipate the role of mobile dry-docks, tenders and repair ships in any Pacific operation. In retrospect, refueling exercises between the wars was nothing less than a dress rehearsal for the 1940's.

The fleet's insatiable oil demand strained U.S. petroleum output. Still, U.S. petroleum firms supplied 99% of the armed services petroleum needs in World War II. Naval reserves set aside accounted for less than one percent. The government oil set aside, dating back to the Taft administration, had proved a tantalizing illusion.

In the last year of war, U.S. domestic oil production leveled off. The Navy now required oil from Venezuela's Lake Maracaibo. To hasten oil delivery from the Caribbean the Navy constructed two pipelines across the Panama Isthmus. The Navy also accorded merchant tanker transit priority comparable to U.S. warships.

In the early months of the war, German U-Boat's sent Harry Sinclair's and new Jersey Standard tankers to the Atlantic bottom. Built in 1917, the tanker Edwin Doheny was designated a floating oil storage ship and dispatched to a Pacific anchorage, giving Edward Doheny the distinction of placing an oil tank farm at Pearl Harbor, oil storage depots on the U.S. west coast; a western domestic pipeline link, Caribbean oil access, Maracaibo oil tankers, and Sevron 10 storage tankers. [55]

Three years after the war, Navy Commodore, W.G. Greenway, testified before Congress. In addressing the question of the navy's oil reserves, his message was unambiguous. The Navy asked to transfer the management of its oil reserves back to the Department of Interior. The rationale would not have surprised Warren Harding, Edwin Denby or Rear Admiral John K. Robison. Commodore Greenway testified that Interior "...could do it better." In 1948, the U.S. Senate turned the request down cold. [56]

Historians have reflected on the multiple facets of the Washington naval treaty system. Samuel Eliot Morison concluded that the Washington Treaty compelled the Navy to become logistically self-sufficient. Edward S. Miller observed that the Washington Treaty stimulated the Navy to pursue logistics innovation. Professor Robert Love noted that the defortification article "spurred" the Navy to search for a fixed base alternative. Professor John Bradley stated that, for the U.S. Submarine Service, at least, the absence of Pacific bases had been a "blessing in disguise." Bevin Alexander added that "...the mobile service force was one of the great innovations in naval history and contributed mightily to America's success in the war against Japan. John Kuehn concluded that the "...mobile base project and its realization in naval building policy and war planning reflected some of the most profound innovation of the interwar period."[57]

Kuehn tied the navy's mobile base project to the Washington Naval Agreement of 1922. Absent a fortified base in the Western Pacific, the Washington treaty inspired the U.S. Navy to embrace the concept of a sustainable fleet afloat.[58]

SOURCING

Pacific War

Congress spent 100 billion dollars on the Navy during the Second World War, an expenditure that, at the time, exceeded the combined assets of AT& T, U.S. Steel and General Motors. The demand for warships, planes, ordnance and supply material swamped government yards. Of necessity, the Navy turned to the nation's industrial sector for assistance and management. Merchant ship conversion became the order of the day. By the end of war, the navy had recycled over 1700 merchant vessels into naval auxiliaries. U.S. industrial yards constructed 5,000 merchant ship hulls, employing Taylor's assembly and modern welding practices.

Rear Admiral Julius A. Furer recalled that the yard industrial manager created in 1922, carried the Navy throughout the Pacific war. In the late 1930's President Roosevelt attempted to soften his reference to U.S. business executives as "economic royalists" and "money changers." Franklin D. Roosevelt now reconfigured U.S. industry as an "Arsenal of Democracy." An arsenal is a government plant. If the President had in mind a government takeover of the nation's plants in the private sector, Congress's ban on Frederick Taylor's management principle would have migrated and stifled U.S. mobilization effort. In 1949, Congress removed its anti Taylor requirement at U.S. navy yards.[59]

The 76th Congress engaged in a programmatic domestic about-face in 1940. U.S. lawmakers now permitted negotiated contracts, replacing competitive bidding requirements. Congress approved letters of intent that expedited plant construction. Congress cast aside its 50:50 building allocation between government and private yards. Congress funded government defense plant construction, subject to private management. Presumably private management was free to engage in modern business practices. Congress agreed to sell war surplus plants to corporations at the end of the war. Congress permitted contract losses to be deducted from total rather than individual contract earnings. Congress adopted a five-year rather than a 20 year amortization schedule for tax purposes. Congress, in short, suspended many of Woodrow Wilson and Franklin Roosevelt's cherished domestic disincentives - for the war's duration.

The Navy now demanded mass production from Newport's torpedo station. That request proved frustrating. In deference to machinist trade unions, Congress had prohibited the most elementary practices of modern plant administration. Unfortunately, while the Newport Station remained wedded a 16th century craft union, Japan's Kure arsenal, a Taylor convert, turned out an oxygen driven long lance 95 torpedo with telling effect. The U.S. Mark XIV was both flawed and obsolete, compounded by a custom-made work process that stifled mass production. A battle crisis often promotes reassessment. As Bureau of Ordnance study put it: "To meet the demands of the program, the Bureau decided to adopt a new policy and consider the torpedo a production item rather than a tool job."[60]

The Navy turned to firms in the nation's industrial sector. American Can, Westinghouse, International Harvester, the Pontiac Division of General Motors, companies—all had adopted facets of Taylor's scientific management practices. In short, the government reverted to firms conversant with assembly line expertise, a decision that stood as a silent rebuke to congress's attempt to act as a trade union shop steward.[61]

The Bureaus of Construction and Repair and Engineering, now the Bureau of Ships converted commercial tankers into naval oilers. That that duality blurred distinctions between commercial and military assets was illustrated by a rail diesel engine, a steam valve, a mobile dry dock, a refrigerated ship, a fleet tug, a Link trainer, a Great Lakes ore boat, an FM radio set, high octane gas, a Mississippi sugar barge, controlled pitch propellers and water fog nozzles. A deconstructed destroyer unveiled gears, wires, circuits, electric motors, turbines, valves. An auto assembly line, a radio factory, an instant coffee plant, a clothing factory, an oil refinery, all were essential to the nation's capital infrastructure. Labor, skilled, unskilled, professional, nonprofessional—male and female lawyers, factory managers, dentists, accountants, investment bankers, expediters, machinists, auto repair mechanics, pharmacists, civil engineers, law enforcement personnel—as reserve personnel lent their talent and expertise to the Navy Department.

By the end of the war, private industry had built two-thirds of the Navy's fighting ships and virtually one hundred percent of all fleet auxiliaries. Indeed, the U.S. had constructed not one but two fleets; a modification of the 1940 combatant ship buildup; and an admixture of ships and landing craft essential to offensive logistics and amphibious operations. At the beginning of the war, the Navy carried 1,000 ships in its inventory. By 1945, that number had grown to 150 million tons.

The range and diversity of ship components overwhelmed the Navy's inventory system. Given its scattered advanced bases across the Pacific, the west coast became a supply bottleneck. The navy solved the problem by reorganizing its shore command structure.

Harding/Denby

Edwin Denby concluded that an experienced, constructor should be in charge of U.S. warship construction at Navy Yards. As ship repair emerged as a top priority in 1944-1945, the Navy's industrial yard manager played a critical role in recycling repaired vessels.

Cognizant of the duality of any commercial asset, Denby, in testimony before Congress, observed that merchant vessels were critical to the fleet's fighting ships. In the 20's Harding's outsourcing endeavors met with minimal success. On the other hand, President Harding promoted an economic incentive system hospitable to environment of risk and capital investment. The 1920's witnessed an expansion of the nation's industrial infrastructure.

On the fiscal side, the administration reduced personal taxes, corporate taxes, and excess profit taxes, generated a budget surplus, reduced the national debt, and promulgated an environment receptive to economic growth. Critics would label the Harding Administration policies toward organized labor as hostile if not retrogressive. What is often forgotten is that real labor wages increased during the decade of the twenties.

MOBILIZATION/CONVERSION

Pacific War

After Poland's capitulation to Germany the European war appeared to take on the traits of a stalemate. In the spring of 1940, German tanks raced through the Ardennes, severing the link between French and British armies. By the end of May, France was on the verge of collapse and Britain was forced to evacuate three hundred thousand members of its Expeditionary Force.

In the summer of 1939, President Roosevelt ordered the Army-Navy Joint Board and the Army-Navy Munitions Board (ANMB) to report to the office of the President. Under the Office of Production Management, William Knudsen's assignment appeared clear enough, negotiate defense contracts, promote factory conversion, expedite war production. Union leaders advised Knudsen to order the automobile industry to cease civilian car production immediately and convert U.S. plants to tanks, trucks, diesel engines. Knudson recalled that it took the Ford Motor Company twelve months to design and alter machine tools for the Model A. Knudsen also observed that converting tools, dies, jigs, fixtures, boring machines was not instantaneous process. It would consume an estimated of 18 months, to say nothing of an additional six months to train employees.

William Knudsen predicted that instant plant conversion would toss skilled employees on the streets. Looking for jobs elsewhere, employees would migrate to other sections of the country, dissipating a firm's skilled labor pool. When the auto plants were ready to ramp up war production, the company's employee's base would all but have evaporated. "Cold turkey" conversion would yield less, rather than more war production.

Residing in a bureaucratic backwater, the ANMB set production priorities and allocated output and materials among and between the armed services. Undersecretary James Forrestal asked Ferdinand Eberstadt, a Wall Street investment banker, to evaluate the past studies of the Munitions Board. After examining the record, Eberstadt wrote Forrestal that it was a "sad commentary" that Congress had neglected two decades of mobilization work and preparation. Forrestal then asked Eberstadt to serve on the Board; Eberstadt turned down the request, then reversed himself after the Japanese Pearl Harbor attack. Eberstadt combined a formidable intellect, limitless energy and gift for incisive decision-making. His stewardship on at the Munitions Board would leave an indelible mark upon the U.S. war effort. [62]

Eberstadt confronted a machine tool shortage, a scarcity of aluminum and copper, and a shortfall in ship parts and components. He resolved the machine tool shortage by locating second hand and unused tools, placing them into a common pool, subject to Munitions Board rationing. Within four months, Eberstadt increased the nation's machine tool supply by 30%. [63]

Eberstadt next addressed a copper and aluminum shortage. He set up a central allocative scheme that distributed the metals on a priority basis. He applied the same technique to components essential in ship construction. Eberstadt installed a rating system that balanced the supply of war production on one side against the demand for war output on the other. The rationing list was so effective that on more than one occasion, the Secretary of the Navy attempted to bypass the Board's precedent rating system. [64]

Concerned that competitive bidding would concentrate production along the U.S. east coast, Eberstadt dispersed contracts geographically. The Bureau of Supplies and Accounts, however, insisted on retaining to a competitive bidding system that had worked well in 1917. By late 1942, the Munitions Board was folded into the War Production Board and Eberstadt would be eased out of office. Historians would have the last word, however. Eberstadt, they said, had shortened the war by twelve months. [65]

Because the President had assigned the Office of Production Management (OPM) responsibility without comparable authority, production shortages beset the nation's 1941 mobilization effort. The President proceeded to improvise a series of regulatory bodies, the National Defense Mediation Board; the Office of Price Administration and Civilian Supply (OPA); the Economic Defense Board; the Office of Defense Transportation; the Office of Emergency Management. Agency jurisdiction inevitably overlapped and generated bureaucratic in-fighting. Disputes peculated to the top, brokered by the President. One historian likened FDR to a spider in the center of in a vast, complex web. In any event, the President spawned duplicative agencies - a propensity that not only enlarged his span of control, but generated its own paper backlog. One historian concluded that, rather than the nation's private sector, it was the President's oval desk that constituted the nation's production bottleneck. [66]

The House and Senate rallied to the nation's war effort as the administration redefined U.S. business as the line behind the front line. Congress began to accept, however reluctant, the management talent and expertise that resided in the nation's private sector An OPA economist, nevertheless, could not resist critiquing a contractor's marketing effort. Learning of the official's tendency to lecture, Congress ruled that of Office of Price Administration personnel had to demonstrate five years of business experience before qualifying for government service. Dr. John Kenneth Galbraith would eventually turn to college teaching. [67]

FDR's proclivity to assign responsibility absent authority continued throughout the early years of the war. Not so General George Marshall. Upon becoming Army Chief of Staff, General Marshall discovered that some 60 officers reported to him directly. Marshall reorganized army so that six officers report to him. To George Marshall, delegation was nothing less than a management imperative. [68]

Lend Lease (1941) disrupted U.S. industrial mobilization and inspired the President to set up the Supply Priorities and Allocation Board (SPAB). Under SPAB, William Knudsen reported to Donald Nelson; under OPM, Donald Nelson reported to Knudsen. By 1941, President had generated so many regulatory czars that Donald Nelson recorded in his diary that "1941 will go down in history, I believe, as the year we almost lost the war before we got into it." What may have been acceptable in an environment of domestic welfare agencies translated into a luxury when time had become a scarce commodity. [69]

After the Japanese Pearl Harbor attack, the War Production Board (WPB) superseded the NDAC, the OPM and SPAB. The WPB attempted to centralize the nation's war effort. Led by Donald Nelson, a former Sears Roebuck vice president, the WPB established production priorities and precedence, while the government mediated labor contract disputes.

In the spring of 1942, U.S. Joint Chiefs of Staff proposed a cross-channel invasion of Europe. Winston Churchill recommended a North Africa invasion instead. Roosevelt sided with the British Prime Minister. The North African invasion required amphibious landing craft, floating assets that were essentially nonexistent. Nelson turned the landing craft problem over the Bethlehem Steel Corporation who promptly evacuated Washington's jammed phone lines and relocated to New York City. The company set up an expediting program. Although the invasion missed its target date of early

November, Operation Torch did take place in the latter half of the month. This was not, however, to be the last experience of production "storming."[70]

Two years later, the invasion of Normandy coincided with the navy's Mariana's offensive in 1944, precipitating another landing craft crisis. The Joint Chiefs of Staff sent the War Production Board its landing craft estimates. For some reason, Nelson questioned the numbers. He flew to London and interviewed General Frederick Morgan, in charge of invasion planning. Morgan verified Nelson's suspicion that the Joint Chiefs had underestimated the Normandy logistic requirement. Nelson sent a telegram to the War Production Board to ramp up production. In the process, the government encouraged contractors to outsource component supply to outside subcontractors. The Navy obviously experienced a quiet transition. A vertically integrated institution now touted the virtues of outsourcing. Although landing craft shortages postponed a southern France landing, the Normandy invasion did take place in June.[71]

Engaged in a two ocean war and short of time, the Navy's shore establishment adopted a subcontracting practice not unlike that employed by the U.S. Maritime Commission. The Commission, in fact, let contracts to construction companies and repair shops along the Ohio, Mississippi, Missouri rivers as well as the Great Lakes. As one officer put it "In all these instances the needs were ultimately filled by private industry with the impetus and inspiration coming from the Marine Corps."[72]

The Reconstruction Finance Corporation (RFC), denying Gulf Industries funding in the early days of the New Deal, financed two-thirds of U.S. war plant expansion. The Defense Plant Corporation, an RFC subsidiary, lent contractors funds to build landing craft, merchant ships, steel plants in Fortuna, California and Genera, Utah; a marine power plant on the west coast, a ship building operation in Tampa, Florida. The Defense Plant Corporation funded factory construction but the government turned over plant operation to private managers. In an ingenious move the U.S. bypassed congress's ban on scientific management applied to navy yards and stations.[73]

Although the Navy anticipated converting hulls into transports, oilers, cargo, refrigerated ships, and ammunition ships, the department was swamped by the volume of its conversion backlog. At first, Navy Yards arrogated the conversion task to themselves. Later, they farmed out the process to private yards—that in turn, outsourced their own vessel requirements. To hasten the conversion process the Navy adopted standard commercial parts and components.[74]

Turbine and boiler manufacturers supplied carrier, cruiser and battleship propulsion machinery. Electric power companies, General Electric, Westinghouse and Allis Chalmers delivered stationary turbines to private electric customers. New York's Consolidated Edison took the step further. Its civilian personnel repaired damaged fleet turbines.[75]

No longer assigned to confront Japanese warships alone, U.S. submarine began to concentrate on Japanese merchant ships. So effective were U.S. submarine tactics that the Japanese fleet was forced to station its ships at Singapore to be near its Indonesian oil supply.[76]

Dual use conversion and time compression continued apace. A Florida swamp boat served as a prototype for an amphibious tank—supplied by Food Machinery Corporation and Borg Warner. A Louisiana rescue craft acted as a model for Marine and infantry personnel landing craft—LCVP; the Higgins Company designed a tank lighter within a period 48 hours that became a standard amphibious asset. A stationary pilot trainer device represented an intriguing dual use application. By 1945, the Link trainer device had instructed some 500,000 Navy, Marine and Army Air Force pilots.[77]

Professor John Craf observed that, prior to the war, washing machines, vacuum cleaners, refrigerators, oil burners and windshield wipers accounted for 90% of the nation's fractional horsepower motors. One B-29 bomber, according to Craf, required more than 300 motors. The automobile industry turned out tanks, aircraft engines and bombers. Louisiana Sugar barges served as oil storage units in the far Pacific; the Navy contracted out some 150 mobile dry-docks from construction companies along the Illinois and Mississippi waterways. Routed through the Panama Canal, Liberty ships towed dry-docks and barges to the western Pacific. Oil tankers delivered fighter aircraft to the pacific. Piper Cub aircraft flew off jerry-rigged LST platforms to serve as artillery spotters. And although Britain had anticipated a family of amphibious ships and craft, the LST, the LSD, LSI, LCI and the AGC, it was U.S. industrial capacity that churned out those assets in scale, quantity, and volume.[78]

The nation's petroleum companies produced high test gasoline to alleviate automobile pre-ignition knocking. High test octane gas accorded allied planes a margin in speed, range and climbing capability. Standard Oil of New Jersey noted that its octane gasoline provided the winning edge to the RAF's Battle of Britain. Caribbean's Aruba oil refinery had supplied the difference.[79]

Prior to 1940, the Navy projected a battleship life of a 15-20 years; an aircraft carrier 15 years; destroyer and submarines,10 years. The war compressed those life cycles. Battle action reports accelerated product innovation in radio, radar and sonar as product life cycles collapsed from years into months. The Grumman Wildcat, a Navy fighter, experienced 200 design changes during its life cycle. Mounted rockets leveraged a plane firing power; proximity fuses and radar fire control enhanced gun accuracy, Combat Information Centers and radar picket destroyers served as an air defense system.[80]

By 1945, the U.S. industrial sector had overcome a host of economic hurdles. Before the Pacific War, Admiral Isoroka Yamamoto had acknowledged as much as he had warned, "...Anyone who has seen auto factories in Detroit and the oil fields of Texas... knows that Japan lacks the national power for a naval race with America." U.S. factory output would overwhelm their Japanese counterpart.[81]

A second U.S. policy was self-inflicted - Federal Reserve Bank's interest rates hobbled capital investment formation in the late 20's and early 30's, compounded by Congressional proclivity to tax, spend, and regulate the economy out of its 1930 malaise. Monetary policy was still somehow perceived as a costless exercise.

Third, the 1939 Army-Navy Munitions Board, recommended that a single administrator, a civilian responsible for coordinating both the military and civilian sectors of the economy. President Roosevelt filed the plan in an office safe. The proposal was not adopted until the spring of 1943 when, under James Byrnes, the administration's War Production and Reconversion Board superseded the War Production Board. Bernard Baruch would later observe that the four year delay[82] "...cost us, unnecessarily, thousands of lives, extra billions and months of tie, largely because FDR was reluctant to delegate authority."[83]

Prior to the war, career naval officers shied away from amphibious operations and logistics assignments, viewed auxiliaries as a career dead end. Most officers preferred a heavy warship assignment.[84]

Filling the gap, reserve officers accepted non-combat posts that permitted "...regular officers to preserve control over operational command." Residing under the Office of Naval Operations, Naval and Marine reservist expanded, supplemented by Naval Reserve Officer Training Programs (NROTC). The Navy introduced accelerated college advancement programs (V-12) and promulgated curriculum geared to naval aviation. (V-5) As one study observed,

> Reservists accounted for lower command assignments in battleships (40%), submarine (70%) and carried escorts (70%). In some cases, reserve officers attained command slots on cargo and transport ships (80%), destroyer escorts (90%) motor torpedo boats (90%) and amphibious craft (43%). By the end of the Pacific War, reservists dominated naval aviation and Marine aviation slots. Members of the merchant marine reserve program commanded fleet transports and cargo vessels as well as running the mobile dry-dock operations of Sevron #10.[85]

The Navy employed reserve officers, experienced in industrial manufacturing, to check, monitor, and expedite arsenal and shipyard output. When production encountered logistics snag on the west coast, a General Motors Vice President of export sales was appointed a Captain and detailed to unravel the inventory pileup.[86]

The war over in 1945, Congress explored legislation restructuring the U.S. defense establishment. Secretary James Forrestal turned to his friend and colleague, Ferdinand Eberstadt. Eberstadt corralled thirty reserve officers and embarked on a crash study summarizing the lessons of the navy's mobilization experience.

Harding/Denby

The mobilization record of the First World War had proved to be a sobering experience. Having campaigned on a platform of peace Woodrow Wilson reversed himself one month into his second term and asked Congress to declare war against Germany and Austria - Hungary. U.S. Army officers informed U.S. industrialist that the war would probably last five years. Before U.S. military production gained traction, the war ended on November, 1918.

Congress held hearings on the nation's mobilization effort and a subsequent the legislation, the Defense Act of 1920, assigned the nation's mobilization mission to the Assistant Secretary of War. Two years later, Secretary of the War, John Weeks, and Secretary of the Navy, Edwin Denby, signed a general letter that set up an Army-Navy Munitions Board (ANMB). The purpose of the board was to coordinate the buying practices of the U.S. military. The ANMB struggled as an institutional orphan as the Navy's Bureau Chiefs distanced themselves from what they regarded as questionable the legal authority of the Assistant Secretary.

During the First World War, Warren Harding, as Senator, had voiced reservations over Wilson's takeover of key U.S. industries. He criticized the administration's handling of the nation's railroad and questioned the need of a government-run shipping company, and voted six times encouraging navy yards to adopt modern business practices. [87]

In the late 1930's the Naval Research Laboratory developed radar equipment. The Tizard committee unveiled a pulsed, cavity magnetron in the fall of 1940. MIT's Radiation Laboratory designed over 150 different radar systems. U.S. radio and electronics firms supplied over one million magnetron tubes, spending over three billion dollars of related radar equipment and apparatus. [88]

Combined with the nation's universities and corporate industrial research, the U.S. was able to take advantage of the forthcoming radio/radar revolution. Dr. Thomas C. Hone put his finger on the issue when he observed;

> ...the electronics industry was strengthened by having to serve a large commercial market. Only the U.S., for example, had the industrial capability of producing high quality magnetrons in the 1940's and only the U.S. had produced frequency modulation or FM radio transmitters and receivers in large quantity by 1940. [89]

And what of Japan's relationship between its navy and civilian scientists? According to David Evans and Mark Prattie, Japanese naval officers so distrusted civilian scientists that they banned them from boarding their ships. In doing so Japan's navy severed a critical technical nexus between customer and supplier.[90]

Intergovernment Coordination: Pacific War

During the war Inter-agency coordination evolved on several levels. Prior to the Pacific war Congress accepted the rule of "mutual command," a cooperative interface between the Army and the Navy. December 7, 1941 exposed the weakness of that protocol. In Hawaii, the Army was assigned to defend the island on lands; the Navy to defend the island at sea. The Commander in Chief of the Pacific fleet, Admiral Husband Kimmel, the Commander of the Army, Walter Short, played golf and participated in the island's social gatherings. Admiral Kimmel, however, reported to CNO Admiral Harold Stark; General Short reported to General George Marshall. Dr. William Atwater later observed, "in many ways it proved to be a tragic choice because it divided authority and allowed a decision in Hawaii where there never existed a clear cut line as to which service had the primary responsibility for air defense."[91]

By October 1944, the Pacific Third Fleet under Admiral William Halsey reported to Admiral Chester Nimitz; the seventh fleet under Admiral Thomas Kinkaid reported to General Douglas MacArthur. The battle of Leyte Gulf has been documented elsewhere. A miscommunication between Halsey and Kinkaid enabled the former to vacate the Philippines beaches leaving the unloading transports essentially unprotected. The Kinkaid/Halsey dispute revealed, once again, the consequences of command disunity.

Congress absolved itself of any responsibility for U.S. military command failure. Rather, congress assigned the fault to the Commander on the spot. Their heads would roll, their reputations destroyed. After the war, Congress legislated a Department of Defense, thereby formalizing command unity. By that legislation the House and Senate acknowledged that Pearl Harbor had been a systemic, rather than an individual command failure.[92]

President Roosevelt preferred to be both his own Navy Secretary and his own Secretary of State. Denied access to the foreign policy implications of the battle front, Cordell Hull was relegated to the role of congressional liaison officer. His successor, Edward Stettinius Jr. was more assertive and anticipated the post-World War II role for the State Department. Indeed, Stettinius accompanied the President to Yalta, advising Roosevelt on matters of Poland, France, the United Nations and the Soviet Union's participation in the Japanese war.[93]

Harding and Denby

Warren Harding, viewing executive reorganization as a matter of high priority, presented to a plan that contemplated a merger of the War and Navy office into a new Department of Defense. Harding thus advocated a system of command unity. The President died before congress took action and apprehensive that their respective services would loosen their individual air force, both John Weeks and Edwin Denby testified against the new proposal.

On the other hand, Weeks and Denby did create an Army-Navy Munitions Board. The ANMB took as its mandate the coordination of industrial mobilization in the event of a national emergency. The War Department then created an Army Industrial College as a vehicle to acquaint officers with the complexity of corporate decision-making. Both John Weeks and Edwin Denby also attempted to involve the State Department in military planning. State, however, begged off and a formal liaison system would await future action.

Summary

In the early stages of the Great Depression, the nation struggled as the stock market imploded, margin calls traumatized investors, factory layoffs accelerated, the nation's output plummeted, unemployment reached double digits. The government's tax receipts fell by 50% and U.S. banks shut their doors. Searching for a culprit for the nation's ills, Congress pointed an accusing finger at the nation's business community. The President improvised a series of economic impediments that handicapped the growth of country's industrial and financial sectors.

By 1940, Congress was forced to reconsider its antibusiness stance. The nation began defense preparations as Britain, now alone, confronted a German jaugernaught. Between 1940 and 1945, relations between Congress and the business community experienced an accommodation of sorts. Congress turned to a "money lender" to resolve a shortage in the nation's tool and died industry; the administration asked an "investment banker" to allocate shortages of steel, cooper and aluminum. The administration called upon on a "merchant of death" to cure a shortage of ship parts and components. The administration solicited a "bankster" to convert bureau military anarchy into a coherent, procurement operation. The administration would turn over its oil reserves to the detested petroleum vice president while a private oil committee recommended the construction of two oil pipelines from the southwest to the U.S. northeast. By the end of the war, Japan had supplied two pounds to each of its combatant; the U.S. delivered four tons to each combatant. The U.S. production record, in short, validated Admiral Yamamoto's worst nightmare.

And what of the structure and content of the U.S. Navy? Adjusting to a world of electronic technology, the Navy adopted a task force command system, perfected amphibious operations, pushed carrier aviation, enlarged a service force, trained naval reservists, and turned the fleet into a formidable mobile force.

Battle historians would continue to assert that the Navy in the 20's had been assaulted by a hostile environment that included budget cuts, inadequate assets, limited personnel, an onerous treaty system, a fickle congress, forfeiture of battleships, a loss of bases west of Hawaii. Naval archives, nevertheless, suggests a different narrative. Despite its handicaps and burden the Navy in the 1920's experienced a tactical, logistic and doctrinal renaissance.

What accounted for the navy's resourcefulness between the two world wars? Two studies suggests an answer: first, Operation's Division of Fleet Training. A Bureau of Ships study put it this way;

> Despite congressional budget limitations, the navy, for twenty years in its program of readiness, worked under scheduled of operation in competitive training and inspection unparalled in any other navy of the world. Fleet problem tactical exercises, amphibious operations with the marines and army aviation, gunnery, engineering, communications were all integrated in a closely packed annual operations schedule. [94]

Second, an administrative history of Operation's Fleet Maintenance Division emphasized the role of the navy's annual fleet exercises.

> The fleet problems served as the grand dress rehearsal for war. Each successive problem showed the steady evolution of the fleet's air arm and from each problem naval aviation lessons learned which enabled it to develop the power and efficiency with which it met its tasks and won victories in World War II. [95]

So subtle and incremental were Harding and Denby's administrative policies that Congress was able to entertain the illusion that twelve years of misdirected economic disincentives, compounded by the funding of anachronistic fleet assets, had somehow permitted the Navy to vanquish the Imperial Japanese Fleet. Put differently, the 1920's provided an arena whereby the U.S. economy experienced an economic revitalization on the domestic front, while permitting the U.S. Navy to engage in institutional learning on the tactical front.

NOTES

1. Robert H. Connery, *The Navy and the Industrial Mobilization in World War II* (Princeton: Princeton University Press, 1951), 160.

2. George C. Dyer, "National Amphibious Landmarks," United States Naval Institute Proceedings, 72, 8, (August, 1966), 57.

3. Eric Larrabee, *Commander-in-Chief: Franklin Delano Roosevelt, His Lieutenants and Their War*, (Annapolis: Naval Institute Press, 1987), 195.

4. Jules Augustus Furer, Administration of the Navy Department in World War I, (Washington, D.C.: naval History Department, 1959).

5. Kent D. Algire, *Major Logistics lessons of World War*, Naval College, Information Service for Officers, (Marine Corps Library, Quantico, VA, February 1951), 36.

6. Historical Section, Office of Naval Operations, Office of the Chief of Naval Operations, (Washington Navy Department Library, ND), 22 (cited as logistics plan). Also Robert Greenhalgh Albion, *Makers of Naval Policy, 1798-1947* (Annapolis: Naval Institute Press, 1980), 316.

7. Furer, 735. U.S. Navy, Bureau of Ships, *An Administrative History of the Bureau of Ships During World War II*, Vol. III, (Washington: Department of the Navy, 1946), 73. [cited as buships study]. Also Henry C. Ferrell Jr., *Claude Swanson of Virginia: A Political Biography* (Lexington: University of Kentucky Press, 1985), 203.

8. Robert C. O'Connell, *Sacred Vessels: The Cult of the Battleship and the Rise of the U.S. Navy* (New York: Oxford University Press, 1991), 316.

9. Stephen N. Roskill, *Naval Policies Between the Wars: Period of Anglo-American Antagonism, 1919-1929* (London: Collins, 1976), 325.

10. J.H. Martin, Geoffrey Bennett, *Pictorial History of Ships* (Secaucus: Chartwell, 1977), 64; also Mark Allen Campbell, "The Influence of Air Power Upon the Evolution of Battle Doctrine in the U.S. Navy, 1921-1941, Master's thesis, University of Massachusetts, Boston, 1992, 46.

11. William H. Garzke, Jr.; Robert O. Pullin, *Battleships: United States Battleships in World War II*, (Annapolis: Naval Institute Press, 1976), on Alaska Class Battlecruisers. "The design and construction of these white elephants was largely the result of President Roosevelt's interest." 76

12. James O. Richardson, *On the Treadmill to Pearl Harbor: The Memoirs of Admiral James O. Richardson* (with Vice Admiral George C. Dyer), Washington: Department of the Navy, 1973): 222.

13. Robert Greenhalph Albion, *Makers of Naval Policy* (Annapolis: Naval Institute Press, 1980), 18-19.

14. Geoffrey Bennett, *Naval Battles of World War II* (New York: David McKay, 1975), 75-75; also H.F. D. Davis, "Building Major Combatant Ships in World War II," 73, 5, U.S. Naval Institute Proceedings, (May 1947), 567.

15. Jacob A. Vander Meulen, *The Politics of Aircraft: Building an American Military Industry* (Lawrence: University Press of Kansas, 1991), 196.

16. Geoffrey Till, "Airpower and the Battleship in the 1920's." Bryan Ranfit, ed., *Technical Change and British Naval Policy, 1860-1939* (London: Hodder and Stoughton, 1977). Finally, Henry M. Dater, "Aviation and Seapower," U.S. Naval Institute Proceedings, 80, 4, (April, 1954), 429.

17. Robert Schaifer, S.D. Heron, *The Development of Airplane Engines and Fuels*, (Cambridge: Harvard University Press, 1950), 11.

18. Eric Larabee, *Commander-in-Chief: Franklin Delano Roosevelt, His Lieutenants, and Their War* (New York: Harper and Row, 1987.

19. Louis Brown, *A Radar History of World War II: Technical and Military Imperatives* (Bristol: Institute of Physics Publishing, 1999), 237. Brown observes "Radar was so new that even the most recent graduates of the naval academy had learned nothing about it, and this made it increasingly the province of reserve officers," 367; also Barrett Tillman, "Coaching the Fighters," 106, 1, U.S. Naval Institute Proceedings, (January, 1980), 41-42; also David C. Evans, Mark R. Peattie, *Kaigun: Strategy, Tactics, and Technology in the Imperial Japanese Navy, 1887-1941* (Annapolis: Naval Institute Press, 1997), 414-415.

20. W. J. Holmes, *Undersea Victory: The Influence of Submarine Operations in the War in the Pacific* (New York: Doubleday, 1966), 41.

21. Mark Allen Campbell, "The Influence of Air Power upon the Evolution of Battle Doctrine in the U.S. Navy, 1921-1941, master's thesis, University of Massachusetts, Boston, 1992, 46.

22. Stephen Peter Rosen, "New Ways of War: Understanding Military Innovation," *International Security*, 13, 1, (Summer 1988): 152.

23. David MacGregor, "Innovation in Naval Warfare in Britain and the United States Between the First and Second World War," Ph.D. dissertation, University of Rochester, 1990, 224.

24. Robert K. Murray, *The Harding Era: Warren G. Harding and his Administration* (Minneapolis: University of Minnesota Press, 1969), 410-411.

25. D. Conley, "The Impact of Technological Change upon Naval Policy," *Royal United Service Institution Journal*, 133, 3. Also NARA, RG 80, General Board letter, No. 420-21, January 16, 1986), Box 62.

26. William Michael McBride, "The Rise and Fall of a Strategic Technology: The American Battleship from Santiago Bay to Pearl Harbor, 1898-1941," Ph.d. dissertation, Johns Hopkins University, 1990, 287, also Merlot, Pusey, *Charles Evans Hughes*, (New York: MacMillan, 1951), Vol. 2, 519.

27. E.B. Potter, "The Command Personality: Some American Naval Officers of World War II," William Geffen, ed., *Command and Commanders in Modern Warfare*: Proceedings of the second military history symposium, 2-3 May, 1968, (Colorado Springs: United States Air Force Academy, 1969), 236.

28. Paul van Cozens, "The Role of Radar in the Pacific Theater during World War Deployment, Acceptance and Effect," master's thesis, San Jose State University, 1993, 114-115.

29. Albert P. Heiner, *Henry J. Kaiser: Western Colossos* (San Francisco: Halo Books, 1991), 149.

30. James Anthony Ginther, Jr., "Keith Barr McCutcheon: Integrating aviation into the United States marine corps." Ph.D. dissertation, Texas Tech University, 1999, 45.

31. Kent Roberts Greenfield, *American Strategy in World War II: A Reconsideration* (Baltimore: Johns Hopkins University Press, 1963), 105-106. Also Jonathan M. House, *Toward Combined Armed Warfare* (Fort Leavenworth: U.S. Army Command and General Staff School, 1984), 131.

32. U.S. Marine Corps, historical amphibious file, HAF 186-190, communications officers school, Quantico, VA., 15 July, 1946, lecture amphibious operations, amphibious force flagship, 6; also NARA RG 38 USS Rocky Mount, Action Report, 10 May 1945, Box 1378.

33. Lawrence Lessing, *Man of High Fidelity: Edward Howard Armstrong* (Philadelphia: J.B. Lippingcott, 1956), 251.

34. Gerald F. Kennedy, "Quantico, VA., Marine Corps kite balloon station and base," 1917, 229; Paolo E. Coletta, ed., United States Navy and Marine Corps Bases, Domestic, (Westport: Greenwood Press, 1985), "during the war, then, the most important schools at Quantico were the Aviation Ground Officers Schools, later the marine air infantry school."

35. Jeter A. Isley, Philip A. Crowl, *The U.S. Marines and Amphibious War* (Princeton: Princeton University Press, 1951), 575.

36. Roger A. Beaumont, *Joint Military Operations: A Short History* (Westport: Greenwood, 1993), 104.

37. J.F.C. Fuller, *The Second World War* (New York: Duel, Sloan and Pierce, 1962), 207.

38. William M. McBride, *Technological Change and the United States Navy: 1865-1945* (Baltimore: Johns Hopkins University Press, 2000), 209.

39. Daniel Yergin, *The Prize the Epic Quest for Oil, Money and Power* (New York: Touchtone, 1993), 317.

40. U.S. Department of the Navy, Administrative History of the 14th Naval District and the Hawaiian Sea Frontier, Vol. II, 16, Navy Department Library, Washington, D.C.

41. Dan Vander Vat, *The Pacific Campaign: World War II, the U.S. Japanese War, 1944-1945* (New York: Simon & Schuster, 1991), 272.

42. U.S. Navy, Bureau of Yards and Docks, *Building the Navy's Bases in World War II* (Washington: GPO, 1947), Vol. II iii.

43. U.S. Naval Administration in World War II, Chief of Naval Operations, Army Navy Petroleum Board and Petroleum Tanker Division, Vo. I, (Washington: Navy Department Library, N.D.), 242.

44. U.S. Naval Administration in World War II, *History of the Naval Research Laboratory*, Vol. 34, (Washington: U.S. Navy Department Library), N.D., 199, on the role of NRL's Radar group engineers. The Officers and Sevron #6. "Traveling with a squadron, they were able to go from ship to ship, sometimes while underway refueling operations and thus help the whole squadron to reach peak performance in a particular equipment." Also, NARA RG 19, Bureau of Ships, Box 645, Action Report, 18 June, 1943, U.S.S. Rocky Mount, Brunei Bay amphibious operation, June 4-12, 1945 The AGC3 "...continues to be somewhat of an auxiliary cargo and fleet carrier." Also radio and radar maintenance staff have been continuously busy repairing equipment of ships in the task group.

45. W.D. Paleston, *The Influence of Sea Power in World War II* (New Haven: Yale University Press, 1947), 90.

46. Captain Ralph K. James, U.S. Fleet maintenance and battle – damage repairs in the Pacific during World War II, Transactions of Northeast Coast Institution of Engineers and Shipbuilders, 67 (July 5, 1951): 357.

47. "The Pacific Sweep," 32, 11, *Fortune* (July 1945): 240.

48. John W. Frey, H. Chandler Ide, *A History of the Petroleum Administration for War: 1941-1945* (Washington: United States Government Printing Office, 1946), 104.

49. U.S. Department of the Navy, *Administrative History, Commander in Chief Pacific Force* (Washington: Naval Historical Center, 1949): 372.

50. NARA, RG 127, History Amphibious Force, #136, Headquarters, Atlantic Amphibious Force, marine Barracks, Quantico, V.A., September 9, 1941, Final Report of first joint training force landing exercises, New River, North Carolina, 4-12 august, 1941. Smith wrote, "when a transport forms a part of an attack force it is a combatant ship, not an auxiliary. Also F.M. Mosley, "The Navy's Floating Dry-Dock," *Marine Engineering and Shipping Review*, 67, 7 (September, 1945): 146.

51. Michael B. Stoff, *Oil, War and American Security*, (New Haven: Yale University Press, 1980): 19; also Harold Ikes, "My Twelve Years with F.D.R.," *Saturday Evening Post*, (July 17, 1948): 101, on Ike's appointment of Ralph Davis. "This appointment was nothing short of inspirational."

52. Michael Evans, *Amphibious Operations* (London: Brassey's, 1990): 26.

53. Uke Visser's Esso UK Tanker's site, http://www.Aukevisser.nl/uk/indes.htm; also *The Manitowoc Herald Times*, November 11, 1943, "...This month [Manitowoc ship building company] completed the tenth submersible nearly two years ahead of schedule."

54. U.S. Naval Administration in World War II: Chief of Naval Operations, Army-Navy Petroleum Board and Petroleum and Tanker Division, (CNO), Department of Navy, Naval Historical Center, Washington, D.C. Vol. 1, 258 (declassified June 4, 1976). John Saroyan, "Anti Fouling Paints – the Fouling Problems," *Naval Engineers Journal*, 80, 4, (August, 1968), 598.

55. U.S. Naval Administration in World War II: Chief of Naval Operations, Army-Navy Petroleum Board and Tanker Division, Vol. 2, Appendix 57: 194 (Unclassified June 4, 1976.]

56. U.S. Congress, House of Representatives, subcommittee on armed services, Petroleum for National Defense, 80 Cong., 2nd sess. May 20, 1948, 386; *New York Times*, May 21, 1948, 35.

57. Samuel Eliot Morison, *The Two Ocean War: A Short History of the United States Navy in the Second World War* (New York: Ballantine, 1972): 21. Edward S. Miller, *War Plan Orange: The U.S. Strategy to Defeat Japan, 1897-1945* (Annapolis: Naval Institute Press, 1991: 234. Robert W. Love, Jr., *History of the U.S. Navy* (New York: Stackpole, 1992): 535. John H. Bradley, *The Second World War: Asia and the Pacific* (Wayne: Avery Publishing, 1989): 34. John T. Kuehn, Agents of Innovation: *The General Board and the Design of the Fleet that Defeated the Japanese Navy* (Annapolis: Crown Publishers, 2002): 378.

58. Saroyan, 598.

59. Julius Augustus Furer, *Administration of the Navy Department in World War II* (Washington: U.S. Department of the Navy, 1959), 526.

60. Buford Rowland, William B. Boyd, *U.S. Navy Bureau of Ordnance in World War II* (Washington: U.S. Department of the Navy, 1953): 126.

61. Robert Kanigel, *The One Best Way: Frederick Winslow Taylor and the Enigma of Efficiency* (New York: Viking, 1997), 488; also William M. Tsutsui, *Manufacturing Ideology: Scientific Management in Twentieth-Century Japan* (Princeton: Princeton University Press, 1988), 30-31.

62. Eberstadt Papers, Seeley G. Mudd manuscript library, Princeton University, report on army and navy munitions board, November 26, 1941.

63. Robert C. Perez, Edward F. Willett, *The Will to Win: A Biography of Ferdinand Eberstadt* (Westport: Greenwood Press, 1989), 2.

64. L.A. Sawyer, W.H. Mitchell, *Victory Ships and Tankers* (Cambridge: Cornell Maritime Press, 1974), 94.

65. Eliot Janeway, "Mobilizing the Economy," *Yale Review*, 59 (1950-1951): 219.

66. Luther Gulick, *Administrative Reflections from World War II* (Tuscaloosa: University of Alabama Press, 1948), 29.

67. *New York Times*, May 28, 1943, 29.

68. Arthur Hadley, *The Straw Giant: Triumph and Failure, America's Armed Forces: A Report from the Field* (New York: Random House, 1986), 44.

69. Townsend Hoopes, Douglas Brinkley, *Driven Patriot: The Life and Times of James Forrestal* (New York: Random House, 1993), 163.

70. George E. Mowrey, *Landing Craft and the War Production Board, April 1942 to May 1944* (Washington: War Production Board, 1946), 27.

71. Peter John Ames, "The Cause and Result of the Landing Craft Deficit in the Mediterranean in 1943," Honors Thesis, Princeton University, March 31, 1967, 14.

72. Andrew C. Clark, *A Cornfield Shipyard* (Mount Vernon: Windmill Publications, 1991), 62.

73. Frank J. Taylor, "Henry Kaiser's Secret Weapon," *Readers' Digest*, 40, 11 (November, 1942): 62.

74. David O. Woodbury, *Battlefronts of Industry: Westinghouse in World War II* (New York: John Wiley, 1948), 3.

75. Buford Rowland, William B. Boyd, *U.S. Navy Bureau of Ordnance in World War II* (Washington: U.S. Government Printing Office, 1953): 127; also, William Mitsutsui, *Manufacturing Management in Twentieth Century Japan* (Princeton: Princeton University Press, 1998), 44-45.

76. Dan Vander Vat, *The Pacific Campaign: The Pacific Campaign: World War II, the Japanese Naval War, 1941-1945* (New York: Simon and Shuster, 1991): 487.

77. *New York Times*, September 9, 1981, B7.

78. John R. Craf, *A Survey of the American Economy, 1940-1946* (New York: North River Press, 1947), 39.

79. Henriett M. Larson, Evelyn A. Knowlton, Charles Popple, *New Horizons 1925-1950, History of Standard Oil Company* (New York: Harper and Row, 1971): 336.

80. Richard Thruelsen, *The Grumman Story* (New York: Praeger, 1976), 182; James W. Hammond, *The Treaty Navy: The Story of the U.S. Naval Service Between the Wars* (Victoria: Trafford, 2001): 186.

81. John Costello, *The Pacific War; 1941-1945* (New York: Quill, 1982), 81. Also Alan Shom, *The Eagle and the Rising Sun: The Japanese-American War, 1941-1943* (New York: Norton, 2004), 271. Yamamoto on his industry, "…American industry is much more developed than ours—and unlike us they have all the oil they want."

82. Duncan S. Ballantine,182, *U.S. Naval Logistics in the Second World War* (Princeton University Press, 1947), 169.

83. Duncan S. Ballantine, *U.S. Naval Logistics in the Second World War* (Princeton: Princeton University Press, 1947), 169.

84. James E. Watters, Walt Johanson, Melchaloupka, *U.S. Naval Reserve: The First 75 Years* (Newport: U.S. Naval War College, 1992), 169.

85. Duncan Ballantine, *U.S. Naval Logistics in the Second World War* (Princeton: Princeton University Press, 1947): 182, 1986; also, *New York Times* December 6, 1945, 29.

86. Duncan Ballantine, *U.S. Naval Logistics in the Second World War* (Princeton: Princeton University Press, 1947), 183-184.

87. H.F. Alderfer, "The Personality and Politics of Warren G. Harding," Ph.D. dissertation, Syracuse University, 1935, 399.

88. Oskar Blumtritt, Hartmut Petzold, William Aspray, ed., *Tracking the History of Radar* (Discataway: IEEE-Rodgers Center for the History of Electrical Engineering, 1994), Walter Kaiser, "The Development of Electron Tubes and of Radar Technology: The Relation of Science and Technology," 229.

89. Thomas C. Hone, "Naval Construction, Surge and Mobilization," *Naval War College Review*, 47, 3, (Summer, 1994): 79.

90. David C. Evan, Mark R. Peattie, *Kaigun: Strategy, Tactics and Technology in the Imperial Japanese Navy, 1887-1941* (Annapolis: Naval Institute Press, 1997), 415. "...the navy's overzealous secrecy prevented civilian scientists and technicians from going aboard warships on which radar sets had been installed and from understanding who the sets actually functioned at sea."

91. William Felix Atwater, "United States Army and Navy Development of John Landing Operations, 1898-1942," Ph.D. dissertation, 1986, Duke University: 113.

92. Samuel Eliot Morison, *The Two Ocean War: A Short History of the United States Navy in the Second World War* (New York: Backbay Books, 1963): 75.

93. S.M. Plokhy, *Yalta: The Price of Peace* (New York: Viking, 2910): 32-33; also John C. Reis, *The Organization and Control of the U.S. Armed Services* (Baltimore: Johns Hopkins, 1964): 59.

94. Bureau of Ships, An Administrative History of the Bureau of Ships During World War II, (Washington: Department of the Navy, 1952, Vol. II.

95. Administrative History, U.S. Naval Administration in World War II, Vol. 36, DCNO (Air) "Aviation in Fleet Exercises," 1911-1939, 231.

Chapter Six

The Silent Strategists

WARREN HARDING

Warren Harding as President and Edwin Denby as Navy Secretary promulgated a set of decisions that embodied elements of modern organizational strategy. At the time the content of those policies would be unappreciated and unrecognized. Subsequent Presidents and Congress would, in fact, amend, reverse or ignore Harding's and Denby's administrative changes. It was the Navy's Trans-Pacific offensive, World War II that conferred a new perspective to the Administration's planning horizon. Consider once again the background of each individual.

Warren Harding was raised in Marion, Ohio, attended college, taught school for a short time, worked at a local newspaper. With his father's help he acquired part ownership of the Marion Star. Ultimately Harding bought out his partner and he took control of the paper. He set type, wrote editorials, solicited advertising, and set up an employee profit sharing plan. His wife managed the paper's back office and displayed a talent in finances and marketing. Over time the paper prospered.

Harding's role as publisher and editor exposed him to the business community in southern Ohio. He served as a director of a Marion bank, a local lumber company, the Marion telephone company. By the time he entered politics, Harding had not only launched a successful business but was conversant with bank loans, depreciation expense, an income statement, an amortization balance sheet. While President in early 1923, Harding sold his newspaper's interest for $550,000 (the equivalent to $5.5 million in today's dollars.)

Harding soon ventured into Ohio politics. He first served in the Ohio legislature, was elected lieutenant governor, made an unsuccessful bid for the governorship. In 1914, Harding was elected to the U.S. Senate, and served until 1920 when he ran for the presidency.

As Senator, Harding sat on the Senate's Naval Affairs Committee and the Commerce Committee. He supported the U.S. entry into the First World War but soon became disenchanted with the direction of Wilson's domestic war policies. He opposed Wilson's takeover of the nation's railroad, telegraph and telephones companies; questioned the government's entrance into the commercial shipping business; voted against legislation that put the Navy into steel production; supported the adoption of scientific management by government arsenals and plants.

Not all of President Harding's naval policies met with success. Although the President came within one vote of overcoming a senate filibuster, Harding was unable to deliver a program to rehabilitate the merchant marine industry. In retrospect, the President might have secured Senator Robert LaFollette's vote had he elected to reinstate a corporate excess profits tax. Or possibly the President should have backed a soldier's bonus in exchange for his maritime subsidy bill. In any case, the President apparently viewed that trade-off as excessive.

Second, Harding set aside Oil Reserve #4, the north slope of Alaska, in the tradition of Presidents' Taft and Wilson. The naval establishment, of course, supported the President's oil set aside. The navy, on the other hand, demonstrated little expertise in petroleum leasing, to say nothing about duplicating the staff and resources of the Interior Department. In permitting the shore establishment to acquire yet another oil reserve Harding encouraged the Navy to move into an arena beyond its knowledge, training and competence. Simply because naval officers equated an oil reserve with Navy Yard did not necessarily translate into petroleum know how. Moreover, a fourth oil reserve encouraged the department to engage in the arcane world of commodity futures. The Navy had more than enough to do trying to figure out its warship construction costs.

The President suffered another liability, the management of his time. Any executive has a thousand and one claims on his energy. A first rule is to husband one's strength, to set priorities and stay with them. Fortunately, the President possessed a cabinet that afforded him the advantage of delegation; Hughes, Dawes, Mellon, Weeks, Denby, Hoover, Lasker, and Theodore Roosevelt Jr. In contrast to Woodrow Wilson's cabinet, Harding's choices embodied an impressive mix of administrative talent and management experience.

One cannot avoid the impression that Warren Harding was overly conscientious. He worked from seven in the morning to midnight, writing to critics, members of congress, political supporters, and job seekers. His speech commitment bordered on the profligate. The president's personal secretary suggested that Harding's long hours compromised his health. [1]

Fourth, Warren Harding appointed a former Senate colleague, Albert Fall, to serve as secretary of the interior. Fall executed the government's oil leases in accordance with the navy's war planning portfolio. While in office, Fall accepted a loan from Edward Doheny, an oil lessee. Fall denied the financial transaction and when later exposed, would serve a year in prison. Historians continue to remind us of Harding's faulty judgment in taking Albert Fall into the President's cabinet. What is sometimes forgotten is that the U.S. Senate, by acclamation, voted to confirm one of their own members, by passing the usual hearing and review process.

On a positive note, the President addressed over a dozen issues that directly or indirectly impinged upon the fortunes of the Navy Department—decisions that ranged from naval aviation to fleet logistics, from dual use products to amphibious warfare. In each case, Warren Harding identified an administrative problem and put forth a remedy that embraced an element of modern organizational doctrine.

First, the Harding administration inherited a six billion dollar government enterprise that operated without a national budget. The President persuaded Congress to pass a Bureau of the Budget (BOB). The budget gave the President management oversight over the content and direction of government expenditures. Put differently, the President, as CEO, possessed a management tool essential to coordinate the resources of the executive branch of government.

Second, the Bureau of the Budget offered a clinical view of the navy's administrative structure. In 1921, for example, BOB suggested that the Navy Department institute a central staff system to effect bureau coordination. As late as 1944, BOB recommended once greater CNO cognizance over the navy's bottom heavy bias.

Third, as a result of the First World War, the administration inherited a twenty-five billion dollar debt. Once the fighting ceased, a naval construction race erupted among former allies, threatening to push U.S. expenditures even higher. By lancing a naval construction boil, the President enabled private investment to expand and deepen the nation's industrial infrastructure—albeit a long term investment in national security.

Fourth, the Harding administration imposed a tonnage ceiling on battleships, while converting two battle cruisers into aircraft carriers. Knowingly or not, the administration set a limit on a weapons system destined to be superseded by carrier aviation. Historians generally concede that the *Lexington* class carriers influenced the design and configuration of the Essex Class Carriers. In a strategic sense, the President's decision anticipated a weapons system by some two decades.

Fifth, Harding set the stage for the 1922 Naval limitation agreement among former allies. Naval historians have been unusually harsh in criticizing the President's policy decisions. No less than fleet Admiral King acknowledged that had Congress's battle cruisers been completed in the 1920's, those heavy warships would have been obsolete by 1940.

Sixth, in deference to civilians' trade unions, Congress embarked on a program of consummate mischief. Although the wages and benefits of civilian employees exceeded their private sector counterpart, Congress accorded Navy Yard exclusive access to ship repair, confined battleship construction to Navy Yards, permitted civilian employees to impose work rules restrictions on the ship forces, banned naval officers from issuing orders to yard trade union workers, limited outsourcing practices. Insulated by tenure, protected by union power, embracing seniority rules, government civilian unions essentially ran government plant operations. As if that was insufficient congress decreed that government arsenals and yards be prevented from adopting the most elementary principles of modern business practices—cost accounting.

Harding confounded congress and trade union opponents. Rather than taking on government unions by frontal assault, the President elected an indirect approach, a naval disarmament treaty. That that program was effective can be seen by Samuel Gompers' plea to permit government plants to diversify into private sector production and services. In retrospect it was not U.S. government arsenals that played a defining role in the Pacific War. Rather it was U.S. firms that adopted the "American Plan" of the 20's that laid the foundation for industrial mobilization in the 1940's.

Seventh, apropos the Washington limitation agreement, President Harding agreed to confine U.S. Pacific bases to the Hawaiian Islands. Deprived of a Pacific base west of Pearl Harbor, the President put in place an incentive system that inspired the Navy to adopt the concept of mobile logistics. Commencing with refueling exercises, the fleet began inching away from a mentality of fixed base dependency, a doctrinal step that would sire Squadrons #6 and #10 in the Pacific war. Institutional impediments can yield resourceful results.

Eighth, the President resisted congressional pressure to fold Army and Navy aviation into an independent, autonomous air force. In opposing Billy Mitchell's vision of unified aviation, the President forestalled an Army takeover of \naval aviation. In the interwar period, the Army Air Service, not unlike Britain's RAF, opposed air-cooled engines, dive bombers, and tactical ground support. Acknowledging its error in 1939, Britain belatedly reconstituted its Naval Air Arm. By then it was a bit late. In the Pacific war, the Royal Navy adopted U.S. planes, escort carriers, fleet logistics, and fleet tactics.

Ninth, the President set in motion advances in carrier aviation, amphibious operations and submarine design. The latter would play no small part in isolating Japan from its oil resources in the late months of the Pacific war. From the perspective of the submarine service, the treaty's non fortification clause proved a "blessing in disguise." Railroad diesels engines enabled U.S. submarine to achieve a 10,000 mile operating range.

Tenth, Congress had scattered naval aviation activities throughout the department's bureau system. In setting up a Bureau of Aeronautics, Harding concentrated naval aviation decision-making within a single organization. A new Bureau Chief, Admiral Moffett, took the lead in training, personnel, procurement and aviation doctrine. Specialization can promote tactical improvisation.

Eleventh, the President also refused to adopt Britain's Air Ministry that folded military and civilian aviation into a single government agency. In divesting commercial from military aviation, Harding permitted market forces to pace and drive the state of the aviation art. Over the long term, commercial technical advances would redound to the benefit of naval aviation.

Twelfth, the President backed Lejeune's commitment to amphibious warfare, a doctrine that not only amended naval tactics, but redefined the mission of surface vessels as well. By the end of the Pacific war, the U.S. Navy and Marine Corps had created a formidable integrated weapon's system.

Thirteenth, President Harding concentrated the U.S. fleet on the Pacific coast and placed the fleet under a single Commander in Chief. Command unity and command accountability go hand in hand.

Fourteenth, the President discouraged government agencies from engaging to duplicative private sector activities. As Senator, Harding had long held reservations concerning the cost, quality and modernity of government arsenals. Harding also recognized that Congress's propensity to pass deficiency budgets was simply a venue to cover exorbitant cost over-runs. In promoting government out-sourcing, Harding employed the private sector to benchmark the performance of the navy's shore establishment.

Fifteenth, President Harding perceived government as an ally rather than a protagonist to private sector institutions. Congress had long subsidized government shipyards and civilian employees. Yet, when the president attempted to do the same for private shipbuilding yards, the U.S. Senate denounced Harding's effort as a corporate raid on the U.S. Treasury. In 1936, Congress legislated the Merchant Marine Act essentially resuscitated Harding's merchant ship program of 1923.

Finally, Harding viewed the executive branch as a hodge-podge of agencies, replicating Congress's incoherent committee system. Congress put the Navy in the oil business while acknowledging that petroleum expertise resided with Interior. The President's Executive Order, permitting Interior to act as agent on behalf of the Navy, was an attempt to match institutional mission with institutional specialization.

EDWIN DENBY

If Warren Harding addressed the broad reach of the executive branch, Edwin Denby concentrated on the structure of the Navy Department. Born and raised in Indiana, Denby lived and worked for the Chinese Imperial Maritime custom service. He returned to the U.S., attended the college, completed a law degree, enlisted in the Spanish-American War and served as gunner's mate aboard the cruiser *USS Yosemite*. The *Yosemite* participated in the Marine assault at Cuba's Guantanamo Bay.

Three of Denby's decisions ill-served the Navy. A first dealt with torpedo production. Denby deferred to the Chief of the Bureau of Ordnance's policy to concentrate torpedo production at its Newport station in Rhode Island, a "make" decision that not only enhanced the navy's shore establishment, it politicized torpedo production. Rhode Island lawmakers pressured the station to hire workers, imposed union shop rules, banned civilian workers from operating two machines, limited research expenditures. In the Pacific War, the Newport Station delivered untested torpedoes to submarines, destroyers, and naval aviation units, two thirds of them duds. It was not until 1943 that submarine officers, over the opposition of the Bureau of Ordnance, diagnosed the multiple flaws of the mark 14 torpedo. By that time, the Navy had begun to outsource its torpedo production to Westinghouse, Pontiac Motor Company, and International Harvestor, firms that had adopted Taylor's scientific management.[2]

Denby, of course, hardly stood alone in sanctioning a government monopoly. One is prompted to ask where were Denby's successors, Curtice Wilbur, Charles F. Adams, Claude Swanson, and Charles Edison. And where were Presidents Coolidge, Hoover and Franklin Roosevelt, the last of whom acknowledged that he preferred to be his own Navy Secretary? The real issue must focus on Congress. Inverting the navy's ends and means, Congress viewed the fleet as serving the shore establishment, then penalized yard management for engaging in cost reduction programs. Not surprisingly, members of Congress called for the removal of Admiral Robert Coontz, the Navy's CNO.

A second policy shortfall replicated the first. Denby chose to limit submarine production to government yards on the premise that the shore establishment could deliver power plants comparable to those supplied by private contractors. By the late twenties, fleet exercises revealed that government propulsion units were as unreliable as Electric Boat's diesels. In a case of lateral thinking, the Navy discovered that domestic railroad diesels might serve as an alternative propulsion unit. During the war, General Motors and Fairbanks Morse delivered light weight diesels in time for the navy's central Pacific offensive, U.S. submarines later imposed a deadly embargo upon the Japanese homeland.

Denby's third decision implied an over reliance on officer advice and naval intelligence. The General Board, the War Plans Division and Bureau Chiefs asserted that Britain had lifted the elevation of its warship guns. On the basis of this information Denby obtained congressional funding to match the Royal Fleet's gun elevation. Britain's Admiralty emphatically denied such an event had taken place and Denby had to return six million dollars to the U.S. treasury. On occasion, a non decision may be more prudent than a hasty one.

Not unlike the President Harding, Edwin Denby proposed remedies that embodied elements of modern management. First, Denby regarded fleet readiness as the conceptual mission over-riding all other departmental activities. To that end, Denby upgraded the War Plans Section within the Operations office, relieving the General Board of that assignment. The decision enhanced Operations' strategic planning status.

Second, Congress had created an organization resistant to inter-bureau coordination. Denby viewed that structure as dysfunctional. As a remedy the Secretary designated Admiral Coontz as the department's first budget officer. He next inserted a War Planning Section into the Bureau of Construction and Repair, and accorded the CNO coordinating authority over fleet repair and alteration. He placed an Operation's staff member on the Army-Navy Munitions Board and designate a Marine to serve on the executive committee of the Army-Navy Joint Board. He assigned the Navy's War Plans Director on a presidential committee promoting fast merchant ships construction. Overtime, incremental moves can erode institutional bulkheads.

Third, Denby employed a series of committees to a remedy to the navy's bottom heavy bureaus - not unlike a practice employed by Alfred P. Sloan. In a strategic sense, Denby attempted to strike a balance between institutional gridlock and decentralized anarchy.

Fourth, by creating a Division of Fleet Training, Denby placed gunnery, engineering, communication and tactical exercises under Operation's control. Through training manuals and tactical publications, the division institutionalized lessons derived from annual fleet maneuvers.

Fifth, Denby approved a matrix command protocol that permitted the Commander-in-Chief to concentrate on fleet training and fleet tactics. Over time the matrix formula would be emulated by carriers, Sevron squadrons, the Marine Corps and Infantry Regimental Combat Teams. Each move invoked Taylor's doctrine of simplicity and Division of Labor.

Sixth, Denby created a Navy Yard Division, assigning the shore establishment to the assistant secretary. Denby's General Order specified a common accounting standard applicable to all Navy yards. In a corporate sense, the Secretary attempted to orient the shore establishment toward responsiveness and cost effectiveness.

Seventh, Denby supported the fleet's refueling at sea during exercises, a practice generalized to include all warships. Blending top down guidance and bottom up training Denby promoted a deductive-inductive learning process.

Eighth, Denby inherited a fleet divided between the Atlantic and Pacific. Captain Alfred Mahan had long contended that a fragmented fleet constituted a weakened fleet. The Secretary repositioned and concentrated on the Battle Fleet toward the Pacific. More important, he encouraged, the entire fleet to participate in annual exercises.

Ninth, Denby appointed Admiral William Moffett as the aviation bureau's first chief. Moffett would hold that position until his death in 1933. Moffett would later be called the father of naval aviation. Leadership and organizational specialization can foster tactical innovation.

Tenth, Denby appointed John A. Lejeune as Marine Corps head marked a first step in advancing the vision of the marine corps. Denby secured and airfield adjacent Quantico, supported a landing site in the Caribbean, sponsored a naval force dedicated to amphibious operations, sustained the tradition of annual landing exercises. Two decades later, amphibious operations would confer upon the fleet a first mover advantage in the Navy's trans-Pacific offensive.

Eleventh, the Marines employed ground liaison parties to assist tactical air support, an early form of a just in time supply chain. Flat organizations encourage tactical improvisation.

Twelfth, Denby executed Josephus Daniels' Pearl Harbor defense plan to place oil storage tanks at Pearl Harbor and both U.S. coasts. To sustain Pearl Harbor as a naval base Assistant Secretary Roosevelt Jr. and CNO Robert Coontz routed U.S. warships to Hawaiian shore facilities, anticipating Pearl Harbor's site as a Pacific advanced base.

Thirteenth, Denby understood the tactical imperative of tenders, repair ships, transports, supply vessels. The Secretary also reiterated that fleet mobility depended upon a modern merchant marine industry.

Fourteenth, Denby ordered commanders to employ ship forces and auxiliary tenders as a device to foster fleet self-reliance. Sevron #10 and Sevron #6 would owe part of their legacy to the replenishment exercises of the interwar period.

Fifteenth, Denby regarded interservice cooperation as a tactical imperative. Under the Secretary's watch the Navy accepted surplus Army machine tools at Pearl Harbor. In turn, the Navy delivered surplus shells and ammunition to the Army. Marine officers attended the Army's tank school and both services participated in the joint amphibious exercises, marking one step toward inter service coordination.

Sixteenth, Denby's resuscitation of the navy's reserve program was nothing if not prescient. During the Pacific war, reserve officers made singular contributions to Navy material acquisitions through their application of

"...modern administrative and business practices." Reserve officers and enlisted personnel compensated for Congress's neglect in funding logistics, auxiliaries and amphibious assets.[3]

Seventeenth, Denby was committed to commercial radio, telegraph and telephone service and production. As the demand for radio sets exploded, consumer purchases broadened an industry base of tubes, handsets, transmitters, receivers, circuits, and crystals. Wireless equipment would, in turn, spawn a network of retail outlets and catalogue stores. By the late 1920's major telephone and radio firms began exploring the feasibility of television. Adopting short wave radio, the Navy's research laboratory eventually would discover fleet and air detection techniques. As one historian put it, nuclear research may have ended the Pacific war, but radar won the war.

Eighteenth, Denby supported the fleet's drive to extent the time a ship had to return to dry dock. The Secretary permitted the Navy department to support the army's chemical division in the development of anti-fouling paint.

Harding and Denby

Both Warren Harding and Denby's paths would converge on two separate occasions, the Tizard Committee and a Unified Defense Department.

The Tizard Mission

A British chemist by training, Henry Tizard, sought to explore the scientific collaboration between Britain and the United States. In the spring of 1940, Tizard suggested that the British government send Dr. Archibald Hill, a physiologist, to determine whether cooperation with the U.S was at all possible. [4]

Hill arrived in North America and met with U.S. officials, academic researchers, as well as the corporate laboratory scientists. In a subsequent letter to Tizard, Hill noted that U.S. officials appeared cordial to Britain's wartime plight. Hill also implied that the U.S. was working on radio detection apparatus. Finally, Hill recommended that Britain reveal its military secrets to America with no conditions attached.[5]

By August 1940, the Tizard Mission unveiled the magnetron transmitter to the American scientists and personnel of the Naval Research Laboratory. Recognizing Britain's technical breakthrough, the U.S. immediately rushed to manufacture microwave radar. Alfred Loomis, head of a U.S. microwave committee, met with corporate and academic leaders. Loomis and parceled out radar's supply components. AT&T's Bell laboratory agreed to supply the magnetron; General Electric proposed to manufacture the necessary magnets; Sperry Corporation accepted the need to produce crystal mixtures; Westinghouse offered to furnish pulse modulators; RCA agreed to produce cathode ray tubes. The month was November, 1940.

One intriguing issue remained. Why did Britain, the inventor of the magnetron, turn to U.S. manufacturing for its supply needs? Several factors account for that decision. First, England, standing alone, anticipated a German invasion in the fall of 1940. If successful, Germany might get its hands on the magnetron. Britain could prevent that possibility by transferring its magnetron research to Canada or Australia. A second thesis argues that the British radio industry, booked to capacity, was simply unable to accommodate the expected growth in radar production. Britain turned to the U.S. for radar output.

A third factor focused on the industrial base of British radio and telephones. In the latter half of the 19th century, the British government permitted private firms to provide telephone service under a Post Office license. Later the British Post Office acquired its rival, bringing it within the Post Office's telephone family. Britain, in short, created a government telephone monopoly. Subsequently, the British Post Office smothered its local telephone management decision-making, stripping regional operations of any real authority. All local decision had to be approved by the Post Master General. [6]

As civil servants, telecom employees tend to be less than responsive to the needs of its residential and business customers. One Parliamentary study suggested that employees regarded themselves as essentially tax collectors. More critically, the Post Office inhibited telecom executives from reinvesting funds into telephone plant and equipment. All revenue surplus had to transfer to the Treasury (Exchequer) Department. The Post Office, in short, constricted the nation's telephone manufacturing base. Telecom officials, to be sure, bridled at tight central control complaining at one point that the Post Office failed to cover telecom employee pensions. [7]

Britain's radio broadcast industry followed a like pattern. Originally, the British Post Office permitted a private consortium to promote wireless listening sets. By the middle 20's, the Post Office cancelled the consortium's radio license, acquired the company's assets, and adopted its name, BBC (British Broadcast Corporation). A crown Corporation now resided under Post Office authority. The BBC elected to fund its program through the imposition of a radio set tax , rejecting commercial advertising revenues. Subscribers who listened to a radio set absent paying a government tax risked prosecution as BBC trucks roamed the streets rooting out illicit listeners. [8]

By contrast, the American Telephone system, a private corporation, took steps to enhance the reliability of its U.S. network. Responsive to consumer and business demand, the American Telephone and Telegraph Company increased its plant and equipment base, as it emphasized telephone quality and reliable phone access.

By the late 20's the disparity between Britain and the U.S. came into sharp relief. U.S. users operated 14.5 million telephones; the U.K. 1.6 million telephones. U.S. telecom firms generated earnings $60 million dollars, the

U.K. $14 million. By the late 30's, U.S. consumers owned 45 million sets listened to 750 broadcast stations. In Britain had 10.5 million sets, listened essentially to the BBC.[9]

When the Second World War erupted, Britain, of necessity, turned to the manufacturing base of U.S. radio and communication suppliers. As an administrative historian put it, "Although radar is a new industry, it is so closely tied in with general electronic production that the existing manufacturers who produced electronics equipment for the telephone, telegraph and radio industries can also, for the most part, supply that for radar."[10]

When questioned after the war how the Naval Research Laboratory (NRL) aided and assisted the Navy, Dr. A. Hoyt Taylor replied that NRL was able convinced the fleet to adopt short war radio technology.[11]

DEPARTMENT OF DEFENSE

Harding and Denby were joined in absentia on a second occasion. In 1947, Congress created a Department of Defense. Two organization reported to the president's a National Security Council. A first included the State Department, Intelligence, and the Armed Services. A second included the National Security Resources Board representing U.S. private sector industry. Congress in 1947 acknowledged that the nation's Industrial section should be placed on par with U.S. intelligence and the armed services.

The plan was virtually a carbon copy of Harding's 1923 Department of Defense proposal. Harding regarded "National Resources," the nation's industrial sector, as critical to the nation's defense as the Army and Navy. Twenty-four years marked the difference between Harding's 1923 defense proposal and Congress's defense 1947 adaptation.

Denby, as noted, later apposed the President's defense reorganization on grounds that the Navy feared the loss of their aviation unit to an independent air force. For the same reason, War Secretary John Weeks joined Denby in opposition. Colonel Billy Mitchell's air vision continued to haunt both armed services. The 1947 legislation did embody Denby/Week's legacy, however, the Army-Navy Munitions Board. Viewed with contempt by the department's bureau chiefs in the 1920's the ANMB emerged as a critical component of the 1947 legislation.

The Harding administration's contribution to the nation's defense resided in its private sector incentives. The administration's strategy was so subtle that Congress was able to entertain the illusion that banning modern business practices in the public sector, combined with harsh economic disincentives on the private sector had somehow provided the formula for naval success in the Pacific. In reality, the nation's private sector not only bailed out Congress's failed policies it did so by adopting the maligned management practices of Frederick Winslow Taylor. In 1949, Congress quietly dropped the ban on arsenal Taylorism. [12]

NOTES

1. George B. Christian Jr., "Warren Gamaliel Harding," *The Current History Magazine*, 18 (September, 1922):907; Sept. Also Harding papers, George B. Christian Sr. papers, roll 249, letter by George F. Christian Jr. October 15, 1939, p. 476.

2. Robert Gannon, *Hellions of the Deep: The Development of American Torpedoes in World War II* (University Park: Pennsylvania State University Press, 1996) 57.

3. Lamar Lee, Jr., "The Navy Advances in Business Management," U.S. Naval Institute Proceedings, 74, 1 (January 1948): 51.

4. Imperial War Museum, archives, paper of Sir Henry Tizard, AFC, FRS 1885-1959, HTT 251; Tizard in favor of technical exchange between the U.K. and U.S. "It must help materially to enlist the help and powerful resources of the American radio manufacturing industry..." (April-May 1940).

5. Hill papers, 18 June 1940, HTT 58, memo of A.V. Hill, proposal for the general interexchange of scientific and technical information between the defense services of Great Britain and the United States.

6. Post Master, Committee on the Post Office, The Bridgeman Report, Post Office archives, London, U.K. 1932, 21.

7. Vicount Wolmer, M.P. *Post Office Report: Its Importance and Practicality* (London: Ivor, Nicholson and Watson, 1932), 217-219. In the 1920s, British Post Office spent £12 million pounds per year on telephone plant; AT&T spent over 300 million on plant; also Alan Clinton, *Post Office Workers: A Trade and Social History* (London: George Allen and Unwin, 1984), 289.

8. Gleason L. Archer, *History of Radio to 1926* (New York: The American Historical Society, 1938), 252; also, George H. Douglas, *The Early Days of Radio Broadcasting* (London: McFarland, 1987), 73.

9. BBC Handbooks, Accounts and Annual Reports, 1927-2022, Microform, No. 1, Hus 384, License income 1928, £871 Thousand pounds; 1934, £1.70 million pounds.

10. Administrative History of World War II, Civilian Agencies of the Federal Government, Microform, War Production Board, Reel 51, #357, Radio and Radar Activities of the War Production Board, October 9, 1945, 8. Also, Correlli Barnett, *The Audit of War: The Illusion and Reality of Britain as a Great Nation* (London: Macmillian, 1986), 175. "After all, modern equipment could have up to 1000 and more components – a tricky task of scheduling for those to whom Critical Path Analysis (to use the modern term) was a novelty: and few Britain civil servants or military men had attended an American business school (there being no British) to learn such mysteries."

11. A. Hoyt Taylor, *The First Twenty Five Years of the Naval Research Laboratory* (Washington: U.S. Navy Department, 1948), 17. "Probably the most important service of the radio division in the early days was the 'selling' of the high-frequency program to the navy and, indirectly, to the radio communication industry."

12. Arthur Herman, *Freedom's Force: How American Business Produced Victory in World War II* (New York: Random House, 2012.)

Chapter Seven

The Destruction of Edwin Denby

On February 11, 1924, the U.S. Senate voted to remove Edwin Denby from the office of Navy Secretary. What forces conspired to seal Denby's fate? Of the multiple factors at work, three coalesced against the Secretary. They include Denby's naval policies; the Teapot Dome scandal; Senator Thomas J. Walsh's antipathy.

DENBY'S MANAGEMENT DECISIONS

Denby and Thomas Walsh were more than political adversaries. They resided at the polar extreme of the political spectrum. Walsh supported trade union aspirations; Denby backed the imperatives of management. Consider again the background of each.

An attorney in Montana, Thomas Walsh had taken on the state's largest corporation, the Anaconda Mining Company. As a plaintiff lawyer, Walsh also represented aggrieved clients against Montana's railroad companies. Following his Senate election in 1912, Walsh emerged as a committed union advocate. He supported legislation that immunized craft unions from the reach of the Sherman Anti-trust Act, he protected unions from the merger provisions of the Clayton Antitrust Act, he voted to ban and prohibit Frederick Taylor's stop watch and incentive wage compensation at government plants and arsenals. Walsh backed legislation that essentially assigned battle heavy ships construction to government yards. Walsh also had family support. The Senator's son-in-law advised Walsh on ways to solicit and retain labor's support. [1]

In negotiating a naval treaty with former U.S. allies, Britain and Japan, the Harding administration agreed to a ceiling on battleship tonnage, a decision that rescinded Congress's 1916 warship authorization. The cessation of dreadnought construction inevitably led to navy yard layoffs, a prospect that prompted Samuel Gompers, AFL President, to suggest that the Navy eliminate its overhead cost and submit bids for private sector work. Gompers also recommended that all U.S. government ship building requirements be performed at navy yards. Gompers thus sought to ban Taylor principles.

Civilian yard employees also opposed the creation of Bureau of the Budget. Civilian yard workers suspected that a budget office rather than Congress might reduce the workload of the navy's shore establishment.

Operations and War Plans supported Denby's order to give the CNO coordination authority over ship repair and alteration. Civilian employees, on the other hand, preferred that Congress exercise that authority exclusively.

Operation's War Plans supported Denby's order to insert an industrial manager into Navy Yard administration. Civilian trade unions, by contrast, suspected that a knowledgeable constructor might identify "soldiering" and resist an unstated operating reflex, "don't give up the ship."

Denby's order to limit the time a warship remained at a Navy Yard obviously enjoyed had presidential backing. Yard employees, on the other hand, interpreted the Secretary's order as placing a ceiling on worker's overtime compensation.

The Secretary's order requesting ship commanders to stock their vessels with extra provisions and spare parts may have been backed by the Division of Fleet Training. From a trade union perspective, however, the order reduced a warship's dependence on a fixed Navy Yard.

The Secretary ordered the fleet to employ ship forces, fleet tenders and supply ships for routine maintenance and repair work, Operations supported the concept of a mobile logistics force. Yard trade unions, by contrast, viewed a fleet auxiliary as a continental base substitute.

Denby's proposal to create a merchant marine reserve may have been seconded by Operation's War Plan Division. The Seafearers Union, not only defined the reserve as a possible anti-union strike force, but yard trade unions opposed any federal help to private shipbuilding yards as a competitive threat to navy yard work.

In a word, Denby's management decisions ran counter to the vested interest of civilian trace union members. In the past government labor unions had not hesitated in flexing their political power. In Denby's fourth bid as a congressman from Michigan's first district, trade unions, specifically, the Telegraphical Union asserted that Denby favored a union open shop. The AFL backed Frank Doremus, a Detroit Councilman, who defeated Denby's 1910 reelection bid. Labor would earn a return on that political investment when Doremus later voted to ban the scientific management practices at navy yards.[2]

Upon taking office, Denby inherited a department riven by a debate between officers who favored central coordination versus officers who advocated a decentralized status quo. The former favored of a more powerful Operations office; the latter favored a powerful Bureau system. The tension between the two escalated during Denby's tenure as Secretary.

Operations backed Denby's appointment of the CNO as the department's budget officer. The Bureau Chiefs, on the other hand, regarded the appointment as limiting their access to Congress appropriations. Operation's War Plans supported the Secretary's creation of an Army Navy Munitions Board. Fearing curtailment in their procurement autonomy, the bureau chiefs insisted that the ANMB had no legal standing. Denby's Navy Yard Division represented an attempt to standardize the accounting across the shore establishment. The material bureaus perceived cost accounting as an infringement upon Bureau independence. Denby's General Order assigning the CNO's coordinating authority over ship repair clearly ran into Bureau opposition. Denby, in effect, had challenged the Navy's third rail - a Bureau-Congressional political nexus.

NAVAL OIL RESERVES

The Department's naval oil reserve controversy replicated the above pattern. Operations, worried about oil leakage through private drainage, favored a leasing plan that met the requirements of its orange war plan. Rear Admiral Clarence Williams, War Plans Director, recommended that Hawaii's tanks should be constructed above rather than below ground in order to save time. Some bureau personnel insisted that petroleum loss was minimal, if not a nonissue. [3]

When President Harding, given Denby's suggestion, permitted Interior's Bureau of the Mines act as an agent on behalf of the Navy, the chief of the Bureau of Engineering, defined oil reserves as a navy yard, and opposed the President's executive order.

What began as an internal controversy among naval factions the oil lease controversy escalated once congress learned of Albert Falls misconduct. By 1924 President Harding had passed from the scene, Calvin Coolidge was President, Fall had left the government. Denby remained as Navy Secretary.

The Senate instinctively turned to Senator Thomas Walsh for guidance on the public land issues. After all, it was Walsh's investigatory work that uncovered the fact that Fall had accepted a loan from Edward Doheny. And Walsh responded. The Senator stated that the President Harding's Executive Order permitting Interior as navy's agent was illegal; that a crime had been committed; that the Navy's oil leasing contract should be cancelled; that Edwin Denby's oil tanks had violated the will of Congress; that Denby should resign by "sundown."

Edwin Denby had solicited legal advice from the navy's Judge Advocate General. The office responded that Congress's June 1920 bill accorded the Navy Secretary total discretion in the use, exchange and storing of reserve oil. Written by Josephus Daniels, the rider was attached to an Army Navy appropriation bill of nearly $1 billion dollars. Neither read, discussed, nor debated, the rider sailed through the Senate. During the debate on Denby's fate as Secretary, Senator Thomas Walsh insisted that since the senate passed Appropriations bill the unanimously by voice vote by the legislation did not require a roll call vote. Senator Walsh then insisted that he had not voted for the Daniels' rider. Few of the Senator's colleagues would join him in asserting that the Senate could legislate without voting.[4]

Nevertheless, the absence of the rider's legislative history concerned Senator Walsh. During the oil reserve debate he asked his staff to do the following:

> As soon as possible look up the legislative history of the paragraph (rider) in the naval appropriations of June 4, 1920 that is important to the Teapot Dome inquiry. Probably it was incorporated as an amendment to the bill as originally drawn. Find out if you can whether it was proposed by the Navy Department. Examine particularly the hearings or the bill so far as they relate to this particular matter and ascertain from them or otherwise who spoke for the Department in reference to the amendment. Examine the debates in both houses if there were any.[5]

The record is silent as to the staff's findings. Senator John Lodge stated on the floor of the Senate that the rider had neither been debated or not discussed.[6]

As the Robinson resolution neared a Senate vote, Senator Lodge, Massachusetts, argued that removing the Navy Secretary without a public hearing was tantamount to "lynch law." Senator Frederick Hale, Maine cautioned that his colleagues that their action would redound to the ultimate shame of this "honorable body." The Senate passed the Robison Revolution.[7]

One item remained on Senator Walsh's agenda, the status of Rear Admiral John K. Robison. Robison had been placed on a promotion list to receive the permanent rank of Rear Admiral. Senator Walsh, insisting that Robison had "filched" navy oil from the public, ordered Robison's name to be removed from the list. Failing that, Thomas Walsh threatened to ban all officer promotions. Secretary Curtis Wilbur sent the Navy's promotion list to President Coolidge. John K. Robison's name was deleted.

In apposing Denby's coordination moves, Congress sided with officers who favored departmental decentralization. Indeed, after disposing of Captain John Robison, the U.S. Senate promoted several officers to flag rank who had apposed oil tank construction on U.S. coasts, Panama and Pearl Harbor.[8]

Two decades later, what contribution did reserve oil make to the navy's Pacific requirements? #4 (Alaska) was abandoned, oil reserve #3, (Wyoming) ran dry after yielding 2 million barrels of crude instead of a projected 130 million barrels. Reserve #2 (Buena Vista) was so checker-boarded by adjacent drilling that the Navy essentially abandoned the reserve. Only Reserve #1, Elk Hull, remained productive (in 1997 the Occidental Petroleum Company purchased Reserve #1). Ninety nine percent of the navy's petroleum resources derived from commercial drilling and private refineries. When, in 1943, the oil consumption of the Pacific Fleet began to exceed U.S. domestic supply, the U.S. turned to Caribbean oil as its oil source, specifically Maracaibo's the Venezuelan, developed by Edward Doheny, Pan American Oil and Transport Company. As one administrative historian commented, "without oil shipped from this area (Curacao and Aruba) the allied campaign in western Europe and the Pacific would have been difficult to the point of impossibility."[9]

At the end of the Pacific war, Fleet Admiral King summed up his view of the navy's Hawaiian investment; "Our foresight in developing Pearl Harbor and our west coast bases has increased immeasurably our ability to carry on the war in the Pacific." The individuals who experienced the political cost of that investment have all but been forgotten. [10]

SENATOR WALSH'S CORRESPONDENCE

A third anti Denby factor was embodied in Senator Walsh's correspondence. On his first day as secretary Denby received a note from Senator Charles Henderson. Both graduates of the University of Michigan Law School, they had stayed in touch over the years. Senator Henderson's letter appended a note that he had received from Senator Thomas Walsh. Walsh wrote that the Chief of the Bureau of Supplies and Accounts had retired and that the Bureau Chief's slot was now open. Senator Walsh recommended that Commander Emmett Gudger, an outstanding naval officer, be appointed to new Bureau Chief. If accepted Commander Gudger would by-pass the rank of captain and attain the rank of Rear Admiral. Henderson asked if Denby could help Walsh.

Denby wrote later that "I had adopted a policy of appointing Bureau Chiefs from captain to admiral lists alone, so Admiral Potter was made chief of the important Bureau of Supplies and Accounts."[11] Admiral Robert Coontz, commenting on Edwin Denby's promotion standard, recalled the following

> Secretary Denby was honest in the manner of his appointment, even going so far to average up a man's marks throughout his career. Beuret, I recalled at 83.8+ and was chosen chief of the Bureau of Construction and Repair though

not even a candidate for the place. In the other case of Gregory the same procedure was followed.

In turning down Commander Emmett Gudger's promotion Edwin Denby had rejected Thomas Walsh's son-in-law. [12]

A coda erupted at the end of Denby's tenure as Secretary. Senator Thomas Walsh requested that the department make available to him the navy's Intelligence Service. Walsh had in mind additional investigatory work. Presumably, the President, as Commander-in-Chief, could make that call. There is no record that President Calvin Coolidge acceded to Walsh's request. [13]

To sum up, Denby's policies and decisions antagonized civilian craft unions; the U.S. Senate accused Denby of violating a law that Congress had passed then proceeded to disown the law; and more important, Denby refused to convert the Navy Department into a patronage mill. Under normal circumstances a beleaguered Secretary might have survived one force of opposition. Three, however, were overpowering. It was also an election year. When Edwin Denby realized that Calvin Coolidge had abandoned him, the former Marine fell on his sword. In the meantime, the U.S. Congress itself chose to manage and coordinate the Navy's shore establishment, plants, arsenals, yards, stations. Congress's management policies would remain undisturbed, until December 7, 1941. [14]

NOTES

1. Thomas J. Walsh papers, Library of Congress, Manuscripts Division, Subject File B, CA 1913-33. Box 208, October 4, 1930. Emmett C. Gudger references a labor paper for Montana. "If you expect the friends of labor to be in the Senate, you must do more than vote for these friends. You must make additional sacrifices of your time and become active."

2. U.S. Congress, House, Amendment to H.R. 14303, 39 Stat, 345, July 6, 1916, 9782, 9791.

3. NARA, RG 80, Director of War Plans to Chief of Naval Operations, Subject: Fuel Oil Storage at Pearl Harbor T.H., 12 November, 1921, C.S. Williams.

4. Thomas James Walsh and Edward Erickson papers, Scrapbook, Library of Congress, Micro #3, #436, Evening Journal, Wilmington, Delaware, March 7, 1924.

5. Thomas James Walsh papers, Manuscript Division, Library of Congress, Box 208, Subject File B, CA 1913-1933. Walsh memo to Staff (ND).

6. U.S. Congressional Record, U.S. Senate, January 31, 1924, 1719; 1735.

7. 1735.

8. Calvin Coolidge 1872-1933, Harvard University, Lamont Library, Letter from Curtiss Wilber (Secretary of the Navy) to President Calvin Coolidge, January 9, 1925. "I am sending over a list of promotions to the United States Senate, omitting the name of Captain J.K. Robison...." Also, *New York Times*, October 27, 1925, 11.

9. U.S Naval Administration of the Second World War, Robert H. Connery, Marion V. Benington, *Curacao Command Headquarters, Commander All Forces*, Vol. 164, 22.

10. United States, Navy Department, Bureau of Ships, An Administrative History of the Bureau of Ships During World War II, first draft narrative, vol. 1, 1952, 86; see also, U.S. Navy administration in World War II, Administrative History of the 14th District, Vol. 2, 16, Navy Department Library, Washington, D.C.

11. Denby papers, Bentley Historical Library, University of Michigan, March 5, 1921, Box 2, Denby's comment on the candidacy of Commander Emmett C. Gudger as Chief of the Bureau of Supplies and Accounts."

12. Edwin Denby papers, Bentley Historical Library, University of Michigan, Box 2, Response to Thomas Walsh interview with magazine of Wall Street, March 15, 1924, 4, (cited as Bentley Library).

13. Robert E. Coontz, *From the Mississippi to the Sea* (Philadelphia: Dorrance, 1930, 409).

14. Bentley Library, memorandum dictated by Secretary Denby to John B. May on Monday, 3 March, 1924, Box 2.

Chapter Eight

The Denby Interval

Navy historians have been singularly harsh in their assessment of Denby's administrative decisions. Some in fact have implied that Denby sowed the seeds to his own downfall. The Secretary's actions were not conducted in a vacuum, however. His tenure in office was preceded by the naval policies of Josephus Daniels on one side. Denby's policies, in turn, anticipated the navy's experience in the Pacific War. Denby thus straddled two periods of time, an interval between the two World Wars.

Figure 1 depicts the Wilson-Daniels naval era, granting that the schematic oversimplifies the complexity of the navy's administrative structure, the operating fleet, the bureau chiefs, yards and stations. The black line represents congress's coordinating authority. The red line the navy's administrative authority. Under the Daniels era, the navy's central administration exercised token control over the fleet and virtually no control over the department's Bureau System. In reserving that authority to itself, Congress had created a bottom heavy Navy Department.

Congress managed the department's ensemble of yards, stations, plants, and factories despite the wishes of yard Commandants, despite the pervasiveness of Taylor practices adopted in the nation's industrial sector, despite of the spread of scientific management practices in Europe and Japan. Instead, Congress penalized U.S. officials and officers who attempted to achieve Navy Yard cost, efficiency and non-productive union work rules. The U.S. Congress prevented the navy's shore establishment from participating in what historians would later term the Second Industrial Revolution of mass production, prefabrication, assembly lines, machine automation, cost accounting, inventory control, shop floor layout. In deference to civilian trade unions, Congress imposed a 16th century guild system on Navy Yards and arsenals, then rationalized its disincentives as industrial democracy and social justice. Beneath this veneer, Congress sanctioned and operated a political machine useful for incumbent reelection.

Figure 3 depicts the navy's administrative structure under Secretary Frank Knox and James Forrestal during World War II. Now in a two ocean war, Congress stepped back from its micromanaging tendencies, and relaxed its control of the navy's administrative machinery, the shore establishment, the operating fleet. Congress permitted the lead admiral to wear three hats; Chief of Naval Operations and Command in Chief of the U.S. Fleet, Joint Chiefs of Staff. As to Bureau cognizance, President Roosevelt informed Admiral King (Cominch) that he could remove any Bureau Chief whose performance be found unsatisfactory.

During the Pacific War, naval coordination transited from congress to the president, the joint chiefs and Cominch. Although congress continued to ban government practices, the military and the U.S. maritime commission encouraged private yards and shops to engage in prefabricated modules, assembly line production and welding practices. Scientific management principles contributed to an unprecedented level of arms production, permitting the U.S. to undertake two theaters of war in addition to supplying Britain and the Soviet Union.

Congress encouraged U.S. contractors to launch nearly 6000 merchant hulls, celebrated the production flow of Higgins landing craft, applauded Kaiser's construction of tankers and escort carriers. That congressional policies were often contradictory. On one hand, congress encouraged private contractor to push the limits of material output while simultaneously denying navy yards from instituting the bare essentials of scientific management.

It is true that Senator Truman's Defense investigatory committee served as Congress's watch dog over U.S. procurement practices during the war. It is noteworthy, however, that the Truman committee did not recommend that Congress rescind its ban on Taylorism, did not recommend a repeal Hoover's Bacon-Davis (prevailing wages), did not suspend the Walsh-Healey act for the duration. Nor did the Truman committee ask congress to reform its own housekeeping assignments that permitted 17 committees to assert jurisdiction over the nation's rubber shortage.

Figure 2 represents Denby's policies in the early 1920's. Denby defined fleet readiness as his first priority. When the Secretary discovered an overlap between Operations and the General Board Denby assigned long term planning to the Board, elevating the stature Operations' War Plan Division.

Denby next concluded that the Chief of Naval Operations exercised minimal authority over the fleet. Denby approved Operation's reconfiguration into a task force command system, accorded the CNO authority over fleet repair and alteration work. The latter move attempted to enhance bureau response to its client, the fleet afloat.

Similarly, Secretary Denby observed that the Bureau Chiefs operated an independent, autonomous, naval fiefdom. Bureau power impeded any real departmental coordination. Both Harding and Denby recognized the latent energy and resourcefulness of the nation's industrial sector. To tap into that resource, Denby cosponsored the formation of an Army-Navy Munitions Board to coordinate the material requirements of the armed services. In the Pacific war asset subcontracting (outsourcing) enabled Congress to circumvent its yard prohibition against Taylorism. Whatever shortfall and frustration Denby experienced as Secretary, his management changes anticipated the needs and requirement of the Pacific War.

Naval historians are prone to dismiss the 1920's as a lost era, a period of naval stagnation, neglect, retrogression. A reading of archives, public and private, suggest a different narrative. While the U.S. embraced the Jazz Age to ease the trauma of a war to end all wars, the U.S. Navy experienced a quiet renaissance. Edwin Denby was a participant in that renaissance. Figure 2 is thus labeled the Denby Interval.

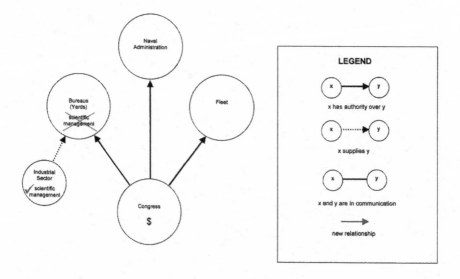

Figure 1.

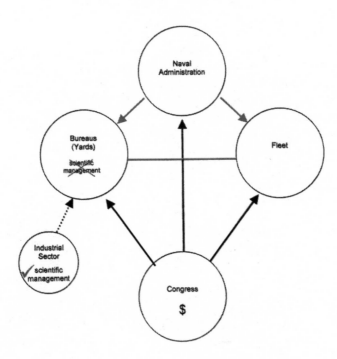

Figure 2.

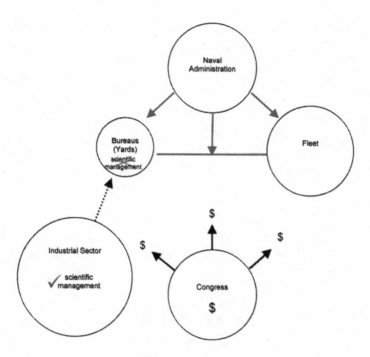

Figure 3.

Selected Bibliography

ARCHIVES – NARA, RG 80

Bureau of Construction and Repair to Chief of Naval Operations via the Major General Commandant of the Marine corps, September 11, 1922.

Commander-in-Chief to Secretary of the Navy, Albert Gleaves, "Observations on Japanese character with special reference to Japan's ambition and her relation in this connection with the United States," 8 December 1920.

Commander-in-Chief, *Fleet Problem No. 4, Report, Blue*, the Value of Having the navy and Military Commanders on the Same Ships: U.S. Naval War College, 4 February 1924.

Development of Channels, Anchorage and Mooring Spaces, Pearl Harbor, from sub-board for the development of Navy Yard Plans to Planning Division, Office of Naval Operations, March 1920.

Director of War Plans to Chief of Naval Operations, Subject: *Fuel Oil Storage at Pearl Harbor, T.H.,* Navy Archives Microfilm, 12 November, 1921.

Fourteen Naval District. U.S. Naval Station, Pearl Harbor from Commandant to Director of Naval Intelligence, subject: visit of I.J.M.S. Yakumo, March 16, 1920.

Memorandum from Chief of Naval Operations to Board on the development of Navy Yard Plans, Mobile Base, June 20, 1922.

Memorandum from Director of War Plans to Chief of Naval Operations, subject, lighters for Landing Expeditionary Force, 10 October, 1922.

Minutes of the Committee of the Secretary of the Navy, June 19, 1922.

Record of the General Board, Naval Archives, Major General Commandant, memo for the General Board, Future Policy of the Marine Corps as influenced by the Conference on Limitations of Armaments, File 432, February 11, 11922.

Secret and Confidential Correspondence of the Office of the Chief of Naval Operations and Office of the Secretary of the Navy, 1919-1927, Edwin Denby, letter to Senator Miles Poindexter, October 15, 1921.

Special Readiness Report, Commander-in-Chief Secret Order, USS Swallow, Bremerton, Washington, 1 January, 1921.

U.S. Office of Naval Operations, Naval Communications, Commander-in-Chief Atlantic Fleet, Amphibious Training Command, Administrative History, World War II, *Navy Department Library*, 1953.

NARA RG 19

Bureau of Construction and Repair, G. Lumbard, Assistant Editor, Marine Engineering and Shipping Review, to Rear Admiral Emory S. Land, Navy Press Release, September 27, 1931.

Bureau of Construction and Repair, General Correspondence, memorandum from Secretary of the Navy to all bureaus and offices, 31 January 1933.

NARA RG 313

U.S. Fleet Base Force, General Correspondence, 1931-1942.

RECORD GROUP 38

Records of the office of the Chief of Naval Operations, Division of Fleet Readiness Training, 1908-1941, General Correspondence, 1914-1941, Vol. I.

Conclusions and Recommendations, Final Report, Army and Navy Joint Exercises, Brig. General J.G. Ord, Commander First Division Task Force and Staff, 1941.

MARINES G 127

Division of Plans and Policies, War Plans Section, Marines, Box 3, *War Plans*, 1923-1933, Memorandum for the assistant to the commandant, from Major General Commandant to Holland M. Smith, U.S.M.C. Headquarters, September 20, 1923.

Division of Plans and Policies, War Plans Section, *War Plans*, Box 3, Mobilization, Marine Corps Archives, 21 July, 1924.

Fleet Problem No. 4, (Box 14), Material Effectiveness, report by Commander-in-Chief, *Marine Corps Archives, Breckinridge Library*, Quantico, VA, February 1924.

HAF, Marine Equipment Board, meetings and reports, Box 1, Marine Archives, Quantico, V.A., memorandum from Commanding of Five, 5th Marines, F.M.F. to Commanding General, Fleet Marine Force, Breckinridge Library, Quantico, VA, 26 March, 1935.

History of Commander Service Force, United States Pacific Fleet, Marine Corps Archives, Marine Corps Education Center, *Breckinridge Library,* Quantico, Virginia, July 13, 1949.

John A. Lejeune, "Amphibious Landing," Lecture, Naval War College, June 1924.

Record of the U.S. Marine Corps Division Plans and Policy, War Plan Section, War Plans 1915-1946, Box 2, Training memorandum for the Major General Commander, from P. Halford, Acting Director, *Marine Corps Archives, Breckinridge Library,* Quantico, VA, August 1923.

U.S. Marine Corps, Division of Plans and Policies, War Plans Section, War Plans, 1915-1946, Box 3, General Correspondence pertaining to war portfolio, memo from Chief of Naval Operations to Major General Commandant, U.S. Marines, *Marine Corps Archives, Breckinridge Library*, Quantico, VA, January 28, 1920.

History Amphibious Forces, *Marine Equipment Board*, Meetings and Reports, Box 1, Breckinridge Library, Marine Archives, Quantico, Virginia, memorandum from 5th Marines to Commanding General, Fleet Marine Force, 26 March 1935, Box 1.

GENERAL BOARD

General Board, Subject File, 1900-1947 GB 420-2 to GB 420-4, Box 63, letter from Secretary of the Navy to Chief of Naval Operations, William D. Leahy, 18 July 1938.

General Board File, Subject File GB 420-2, Box 63 memorandum from a Director of Shore Establishments (C.W. Fisher) to General Board, 10 May 1940.

Hearings, General Board, Air Warfare, naval War College, June 26, 1922.

National Recovery Administration, RG 9.

National Recovery Administration, transcript of proceedings, Shipbuilding Code hearings, first day, Statement of Rear Admiral E.S. Land, Chief, Bureau of Construction and Repair, U.S. Navy, Box 193, July 19, 1933.

National Recovery Administration, Consolidated Approved Code Industry file, Shipbuilding, Transcript of Proceedings, Shipbuilding code hearings, Box 5250, July 21, 1933.

NAVY DEPARTMENT LIBRARY

Operational Achieves, U.S. Naval Administration in World War II.

Operational Archives, United States Naval Administration in World War II, Deputy Chief of Naval Operations, "Procurement of Aircraft," 1907-1939, Vol. 17, Operational Archives, "Aviation in Fleet Exercises," 1911-1939, Vol. 16.

Operational Archives, "The Logistics of Fleet Readiness; Fleet Maintenance Division in World War II," Vol. 20.

Operational Archives, "History of the Naval Research Laboratory," Vol. 134.

Operational Archives, "The Control of Naval Logistics," Assistant to CNO for Material op. 23, Fleet Maintenance.

Operational Archives, Procurement of Organization Under the Chief of Naval Operations, the History of op. 24, Vol. 22.

U.S. Naval Administration in World War II, Commander in chief, Atlantic Fleet, Vol. I, 1953.

Bureau of Ships, *An Administrative History of the Bureau of Ships during World War II*, Washington, DC, Department of the Navy, 1952, Vol. II.

U.S. CONGRESS

CIS, *Index to U.S. Executive Branch Documents*, 1920-1932, Bethesda, MD: Congressional Information Service, 2001.

Congressional Record, March 21, 1916.

Congressional Record, January 31, 1924.

Office of the Secretary of Defense Weapons Systems Evaluation Group, *Operational Experience of Fast Carriers Task Forces in World War II*, (Washington, DC: U.S. Navy), 1951.

U.S. Congress, House. Subcommittee, Appropriations Committee, *Naval Appropriations Bill for 1922*, 66th Cong., 3rd sess., 1921.

U.S. Congress, House, Naval Affairs Committee, *Hearings on Appropriations, Fiscal 1923*, 66th Cong. 1st sess., 1922.

U.S. Congress, House. Subcommittee on Yards on Docks, Naval Affairs, *Authorizing the acquisition of certain sites for Naval Air,* Statues H.R. 11983, 67th Cong. 2nd sess., 1922.

U.S. Congress, Senate. Committee on Commerce, *To Amend Merchant Marine Act of 1920*, Hearing before the commerce committee, 66th Cong., 1st sess., 1922.

U.S. Congress, Senate. Committee on Public Lands and Survey, *Leases Upon Naval Oil Reserves*, Hearings before the Committee, 66th Cong. 1st sess., 1923.

U.S. Congress, House. Committee on Naval Affairs, *Sundry Legislation Affect in the Naval Establishment*, 67th Cong., 2nd-4th sess., 1923.

U.S. Congress, Senate. "Leases Upon Naval Oil Reserves," Committee on Public Lands and Surveys, October 25, 1923.

U.S. Congress, *Joint Committee on the Reorganization of Government Departments, Reorganization of the Executive Departments*, 67th Cong., 4th sess., 1923.

U.S. Congress, House. Committee on Appropriation, *Navy Department Appropriations Bill*, 1924, 67th Cong., 4th sess., 1923.

U.S. Congress, Senate. Committee on Public Lands and Survey, *Leases Upon Naval Oil Reserves*, Hearings before the Committee, 66th Cong. 1st sess., 1923.

U.S. Congress, House. Subcommittee of House Committee on Appropriations, *Naval Department Appropriations Bill*, 1925, 68th Cong., 1st sess., 1924.

U.S. Congress, House. *War Policies Commission*, Message from the President of the United States, 72nd Cong. 1st sess., 1931.

U.S. Congress, House, *War Policies Commission*, 72nd Cong. 1st sess., December 1931.

U.S. Congress, House. Committee on Naval Affairs, *Sundry Legislation Affecting the Naval Establishment,* 64th Cong., 1st sess., 1936.

U.S. Congress, Senate, Special Subcommittee Investigating the Munitions Industry, *Munitions Industry*, 64th Congress, 2nd Session, Feb. 6, 1936.

U.S. Congress, Senate, Special Committee Investigating the Munitions Industry, 64th Cong., 2nd sess. February, 1936.

U.S. Congress, Temporary National Economic Committee, *Investigation of Concentration of Economic Power*, 76th Cong. 3rd Session, 1940.

U.S. Congress, Senate. Session, *History of Naval Petroleum Reserves*, Prepared by the Navy Department at the request of David I. Walsh, 78th Cong., 2nd. sess., May 2, 1944.

U.S. Navy, Naval Administration, Selected documents on Navy Department Organization: 1915-1940. *Naval War College Library*, 1945.

NAVAL HISTORICAL COLLECTION

Coontz, Robert E., Rear Admiral, *Logistics*, Naval War College, U.S. Naval War College, October 7, 1926.

Earle, Ralph, *Landing Operations of the Control Force*, guest lecture, U.S. Naval War College, 11 December 1922.

Gregory, L.E., Admiral, *Lecture to the class of 1925*, *U.S. Naval War College*, U.S. Naval War College, 15 March 1925.

BRITISH ARCHIVES AND BRITISH LIBRARY

Alan Clinton, *Post Office Workers: A Trade and Social History* (London: George Allen, Unwin, 1984).

BBC First Annual Report, 1927, (London: Her Majesty's Stationary Office, 1928)

BBC Handbook, Accounts and Annual Reports, 1927-2003, No.1, HUS 384, 546, 981.

Imperial War Museum, Henry Tizard Papers, HTT 228, 251, 252, 253, 256, 268, 58, 57.

Imperial War Museum Archives, The Tizard Papers, London.

The Bridgeman Report, Postmaster General, Committee of Inquiry on the Post Office, 1932, British Post Office Archives, London, U.K

Viscount Wolmer, *Post Office Reform: It's Importance and Practicality* (London: Ivor, Nicholson, Watson, 1923).

PERSONAL PAPERS

Coolidge, Calvin, Personal Papers, Lamont Library, Harvard University, Roll #21.

Denby, Edwin, Detroit Public Library, Burton Historical Collection, Box 5, letter from Josephus Daniels to Senator Carroll S. Page, Chairman, Senate Naval Committee, U.S. Senate Washington, DC: 21, April 1920.

Denby, Edwin, Detroit Public Library, Burton Historical Collection, letter from Josephus Daniels to Representative Lemuel D. Padgett, House of Representatives, Washington, DC: May 21, 1920.

Denby, Edwin, Detroit Public Library, Burton Historical Collection, Box 5 "History of the Naval Petroleum Reserves as shown by the Records of the Navy Department," Navy Department, Washington, DC: 16 May 1922.

Denby, Edwin, Bentley Historical Library, University of Michigan, Ann Arbor, Michigan, Memorandum dictated by Secretary Denby to John B. May, 3 March, 1924.

Denby, Edwin, Detroit Public Library, Burton Historical Collection, Box #5, memorandum for the Secretary of the Navy regarding my recollections as to the naval oil reserves; 6 March, 1924.

Denby, Edwin, Detroit Public Library, Burton History Collection, Box 6, Robison Deposition, United States v. Mammoth Oil Company, et.al. Equity No. 1431, District Court, Wyoming, July 21, 1924.

Denby, Edwin, Detroit Public Library, Burton Historical Collection, Detroit, Michigan, Box 4, File 495, "speeches and newspaper articles." No date.

Eberstadt, Ferdinand, Seeley G. Mudd Manuscript Library, Princeton University.

Eitel-McCullough, Petitioner V. Commissioner of Internal Revenue, Respondent. Reports of the Tax Court of the United States, Vol 9. December 17, 1947.

Harding, Warren G., Dartmouth College, (1865-1923) Warren G. Harding Papers, Microfilm edition Columbus, OH: Ohio Historical Society, 1969.

Kettering, Charles, Kettering University, GMI Alumni Foundation Collection: Flint, Michigan. (ND)

Weeks, John W., Dartmouth College, Box 19, Secretary of War, Graduation Exercises of the Army Industrial College, Feb. 2, 1925.

BOOKS

Albion, Robert Greenhalah, *Makers of Naval Policy: 1798-1947*, Annapolis, MD: Naval Institute Press, 1980.

Adams, Henry A. *Witness to Power: The Life of Fleet Admiral William D. Leahy*, Annapolis, MD: Naval Institute Press, 1985.

Aitken, Hugh G.J, *The Continuous Wave: Technology and American Radio, 1900-1932*, Princeton: Princeton University Press, 1985.

Aldcroft, Derek H., *The Inter-War Economy: Britain, 1919-1939*, New York: Columbia University Press, 1970.

Allen, Frederick Lewis, *Only Yesterday*, New York, N.Y.: Harper & Row, 1957.

Allison, David Kite, *New Eye for the Navy: The Origin of Radar at the Naval Research Laboratory,* Washington: Naval Research Laboratory, 1981.

Alter, Jonathan, *Defining the Moment: FDR's Hundred Days and the Triumph of Hope*, New York, NY: Simon and Schuster, 2006.

Amato, Ivan, *Pushing the Horizon: Seventy Five Years of High Stakes Science and Technology at the Naval Research Laboratory* (Washington: Naval Research Laboratory, 1998).

Anderson, Robert Earle, *The Merchant Marine and World Frontiers*, Westport, CT.: Greenwood Press, 1978.

Archer, Gleason L., *History of Radio to 1926*, New York, NY: American Historical Society, 1938.

Ballantine, Duncan S., *U.S. Naval Logistics in the Second World War*, Princeton, NJ: Princeton University Press, 1947.

Barnett, Correlli, *Engage the Enemy More Closely: The Royal Navy in the Second World War* (New York: W.W. Norton, 1991).

Barnett, Correlli, *The Audit of War: The Illusion and Reality of Britain as a Great Nation*, London: Papermac, 1986.

Bartlett, Merrill L., ed. *Assault from the Sea: Essays on the History of Naval Warfare*, Annapolis, MD: Naval Institute Press, 1983.

Bartlett, Richard, *The World of Ham Radio, 1901-1950*, Jefferson: MacFarland, 2007.

Barnouw, Erik, *A Tower of Babel: A History of Broadcasting in the United States, Vol. 1 to 1933*, New York, Oxford University Press, 1966,

Beasley, Norman, *Knudsen: A Biography*, New York, NY: McGraw Hill, 1947.

Beaumont, Roger A., *Joint Military Operations: A Short History*, Westport, CT: Greenwood Press, 1993.

Bennett, Geoffry, *Naval Battles of World War II*, New York: David McKay, 1975.

Beuerchen, Alan, "From Radio to Radar," W. Murray, Alan Millett, *Military Innovation in the Interwar Period*, Cambridge: Cambridge University Press, 1998.

Biley, Kenneth, *The General: David Sarnoff and the Rise of the Communications Industry*, (New York: Harper and Row, 1988).

Bilstein, Roger E., *The American Aerospace Industry: From Workshop to Global Enterprise*, New York, NY: Twain, 1960.

Blum, Alberta , "Birth and Death of the M-Day Plan," Harold Stein, ed., *American Civil-Military Decisions*, Tuscaloosa: University of Alabama Press, 1963), 68.

Blum, John Morton, *V was for Victory: Politics and American Culture During World War II*, New York, NY: Harcourt Brace Jovanovich, 1976.

Blumtritt, Oskar, Hartmut Petzold, William Asprey, *Tracking the History of Radar*, Piscataway IEEE- Rudgers Center for History of Electrical Engineering, 1994.

Boslaugh, David L., *When Computers Went to Sea: Digitization of the United States Navy*, Washington: The Society of Naval Architects and marine Engineering, 1999.

Bowen, E.G., *Radar Days*, Bristol: Institute of Physics Publishing, 1987.

Bowen, Harold G., *Ships Machinery and Mossbacks: The Autobiography of a Naval Engineer*, Princeton, NJ: Princeton University Press, 1954.

Bowman, Waldo G. et.al., *Bulldozers Come First: The Story of U.S. War Construction I Foreign Lands*, New York, NY: McGraw-Hill, 1944.

Boyd, Carl and Yoshida, Akihiko, *The Japanese Submarine Force and World War II*, Annapolis, MD: Naval Institute Press, 1995.

Bradford, James C., ed., *Admirals of the New Steel Navy: Makers of the American Naval Tradition 1880-1930*, Annapolis, MD: Naval Institute Press, 1990.

Bradley, John H., *The Second World War: Asia and the Pacific*, Wayne, NJ: Avery Publishing, 1989.

Brayard, Frank O., *By Their Works Ye Shall Know Them: The Life and Ships of William Francis Gibbs, 1886-1967*. New York, NY: Gibbs and Cox, 1968.

Briggs, Asa, *The Birth of Broadcasting*, (London: Oxford University Press, 1961).

Broehl, Wayne G., *Precision Valley: The Machine Tool Companies in Sprinting, Vermont*, Englewood Cliffs, NJ: Prentice Hall, 1959.

Brown, Louis, *A Radar History of World War II: Technical and Military Imperatives* (Philadelphia: Institute of Physics Publishing, 1999).

Buderi, Robert, *The Invention that Changed the World*, New York: Simon and Shuster, 1996.

Buell, Thomas B., *Master of Sea Power: A Biography of Fleet Admiral Ernest J. King*, Boston, MA: Little Brown, 1980.

Burns, Arthur E. III, *The Origen and Development of U.S. Marine Corps Tank Units: 1923-1945*, Marine Corps Command and Staff College, Quantico, VA, 1977.

Burns, Russell, ed. *Radar Development to 1945*, London: Peter Peregrinus, 1988.

Buxton, Neil K., Derek H. Aldcroft, ed. *British Industry Between the Wars*, London: Scolar, 1979.

Clark, Ronald W., *The Rise of the Boffins*, London: Phoenix House, 1962.

Clarricoats, John, *World at Their Fingertips*, London: Radio Society of Great Britain, 1967.

Coase, Ronald H., *British Broadcasting*, London: Longmans, Green, 1950.

Coles, Harry L., ed., *Total War and Cold War: Problems in Civilian Control of the Military*, Columbus: Ohio State University Press, 1962.

Coletta, Paolo E., *Admiral Bradley A. Fiske and the American Navy*, Lawrence: Regents Press of Kansas, 1979.

Coletta, Paolo E., *The American Naval Heritage*, New York, NY: University Press of America, 1987.

Coletta, Paola E., ed., *United States Navy and Marine Bases, Domestic*, Westport, CT: Greenwood, 1985.

Connery, Robert H., *The Navy and the Industry Mobilization in World War II*, Princeton, NJ: Princeton University Press, 1951.

Cronon, E. David, ed., *The Cabinet Diaries of Josephus Daniels*, Lincoln, NE: University of Nebraska Press, 1963.

Cruikshank, Jeffrey L., Schultz, Arthur W., *The Man Who Sold America* (Boston: Harvard Business Review Press, 2010).

Daniels, Josephus, *The Wilson Era: Years of War and After 1917-1923*. Chapel Hill, NC: University of North Carolina Press, 1946.

Davidson, Joel R., *The Unsinkable Fleet: The Politics of U.S. Navy Expansion in World War II*, Annapolis, MD: Naval Institute Press, 1996.

Davis, George T., *A Navy Second to None: The Development of Modern American Naval Policy*, Westport: Greenwood Press, 1940.

Davis, Vincent, *The Politics of Innovation: Patterns in Navy Cases*, (Denver: University of Denver, 1967), 47-48.

Dawes, Charles G., *The First Year of the Budget of the United States*, New York, NY: Harper, 1923.

Dean, John W., *Warren G. Harding*, New York: NY: Henry Holt, 2004.

DeSoto, Clinton B., *Two Hundred Meters and Down*, West Hartford: The American Radio Relay League, 1936.

Devereux, Tony, *Messager Gods of Battle: Radio, Radar, Sonar, The Story of Electronics in War*, (London: Brassey's, 1991).

Dillon, Mary Earhart, *Wendell Wilkie, 1892-1944*, New York, NY: J.B. Lippincott, 1952.

Dixon, Edgar B., Franklin Roosevelt, Foreign Affair, 1899-1922, (Cambridge: Belknap Press, 1969).

Douglas, Susan, *Inventing American Broadcasting, 1899-1922*, (Johns Hopkins University Press, 1987), 1909-1949, College Station: Texas A & M University, 1991).

Douglass, George H., *The Early Days of Radio Broadcasting*, Jefferson: McFarland, 1987.

Dunlap Jr., Orrin E., *The Story of Radio*, New York: Dial Press, 1935.

Eccles, Henry C., *Logistics in the National Defense*, Mechanicsburg, PA: Stackpole, 1959.

Evans, David C., Prattie, Mark R., *Kaigun: Strategy, Tactics, and Technology in the Imperial Japanese Navy, 1887-1941*, (Annapolis: Naval Institute Press, 1997).

Faulkner, Harold, *From Versailles to the New Deal*, New Haven, CT: Yale University Press, 1950.

Felker, Craig C., *Testing America Sea Power: U.S. Navy Strategic Exercises, 1923-1940*, College Station, TX: Texas A & M University, 2007.

Fearon, Peter, *War, Prosperity and Depression: The U.S. Economy 1917-1945*, Lawrence, KS: University Press of Kansas, 1987.

Ferrell, Henry C., *Claude A. Swanson of Virginia: A Political Biography*, Lexington, KY: The University Press of Kentucky, 1985.

Fisher, David E., *A Race of the Edge of Time: Radar-The Decisive Weapon of World War II*, New York: McGraw-Hill, 1988.

Fleming, Thomas J., *The New Dealers War: Franklin D. Roosevelt and the War within the World War II*, New York, NY: Basic Books, 2001.

Forbes, John Douglas, *Stettinius Sr.: A Portrait of a Morgan Partner*, Charlottesville, VA: University Press of Virginia, 1974.

Frederick, Lane C., *Ships for Victory*, Baltimore, MD: Johns Hopkins University Press, 1951.

Friedel, Frank, *Franklin D. Roosevelt: The Apprenticeship*, Boston, MA: Little Brown, 1952.

Friedman, Norman, *Naval Radar*, Annapolis: Naval Institute Press, 1981.

Friedman, Norman, *Network-Centric Warfare: How Navies Learned to Fight Smarter Through Three World Wars*, (Annapolis: Naval Institute Press, 2009).

Friedman, Norman, *U.S. Naval Weapons*, Annapolis: Naval Institute Press.

Fuller, J.F.C., *The Second World War*, New York, NY: Duel, Sloan, Pierce, 1962.

Furer, Julius Augustus, *Administration of the Navy Department in World War II*, Naval History Division, Washington, DC: 1959.

Gebhard, Louis A., *Evolution of Naval Radio- Electronics and Contributions of the Naval Research Laboratory*, Washington, Naval Research Laboratory, 1979.

George, James L., *History of Warships: From Ancient Times to the Twenty-First Century*, London, UK: Constable, 1998.

Gordon, John Steel, *An Empire of Wealth, the Epic History of American Economic Power*, New York, NY: Harper Collins, 2004.

Greenfield, Kent Robert, *Army Ground Forces and the Air Ground Battle Team Including Organic Light Aviation*, Washington, DC: Historical Section, Army Ground Forces, 1948.

Groupman, Alan, ed., *The Big 'L': American Logistics in World War II*, Washington, DC: National Defense University Press, 1997.

Guerlac, Henry E., *Radar in World War II*, Los Angeles: Tomash, 1987.

Gulick, Luther, *Administrative Reflections from World War II*, Tuscaloosa, AL: University of Alabama Press, 1948.

Hacker, Louis M., *The Course of American Economic Growth and Development*, New York, NY: John Wiley, 1970.

Hadley, Arthur T., *The Straw Giant: Triumph and Failure, America's Armed Forces: A Report from the Field*, New York, NY: Random House, 1986.

Hamilton, John, *War at Sea, 1939-1945*, (New York: Blandford Press, 1986).

Hammond, James W., *The Treaty Navy: The Story of the Naval Service Between the Wars*, Victoria, Canada: Wesley Press, 2001.

Harrison, Michael, *Mulberry: The Return in Triumph*, London, UK: W.H. Allen, 1965.

Hartcup, Guy, *The Challenge of War: Britain's Scientific and Engineering Contributions of World War Two*, New York: Taplinger, 1970.

Hartcup, Guy, *The Effect of Science on the Second World War,* London: MacMillan, 2000.

Hawley, Ellis W., *The New Deal and the Problem of Monopoly*, Princeton, NJ: Princeton University Press, 1966.

Head, Sidney W., *Broadcasting in America: A Survey of Television and Radio*, Boston: Houghton Mifflin, 1956.

Heindl, Robert D., *Soldiers of the Sea: The United States Marine Corps 1775-1977*, Annapolis, MD: Naval Institute Press, 1962.

Heiner, Albert P., *Henry J. Kaiser: Western Colossus*, San Francisco, CA: Halo Books, 1991.

Hezlet, Arthur, *Electronics and Sea Power*, New York: Stein and Day, 1975.

Higgs, Robert, *Crises and Leviathan: Critical Episodes in the Growth of American Government*, New York, NY: Oxford University Press, 1987.

Hoek, Susan van, Marion Layton Link, *From Sky to Sea: The Story of Edwin A. Link*, Flagstaff, AZ: Best, 1983.

Holley, Irvin Brinton, Buying Aircraft: *Material Procurement for the Army Air Forces*, Washington, DC: Department of the Army, 1964.

Hone, Thomas, Friedman, Norman, and Manedeles, Mark. D, *American and British Aircraft Carrier Development 1919-1941*, Annapolis, MD: Naval Institute Press, 1999.

Hone, Thomas C., and Hone, Trent, *Battleline: The United States Navy, 1919-1939*, Annapolis, MD: Naval Institute Press, 2006.

Hopkins, A.D., Evans, K.J. *The First 100: Portraits of Man and Women who Shaped Las Vegas*, (Las Vegas: Huntington Press, 1999); Part I, "James Scrugham: Governor on Wheels."

Hornby, William, *Factories and Plants*, London: Longmans, Green, 1958.

Howett, L.S., *History of Communications-Electronics in the United States Navy*, (Washington: Bureau of Ships, 1963). *James Forrestal*, New York, NY: Random House, 1993.

Hughes, Wayne P., *Fleet Tactics: Theory and Practice*, Annapolis, MD: Naval Institute Press, 1986.

Hyland, L.A. "Pat", *Call Me Pat*, New York: Donning, 1994.

Irwin, Douglas A., *The Smoot-Hawley Tariff and the Great Depression*, draft manuscript, April 5, 2010.

Isley, Jeter A., and Crowl, Philip A., *The U.S. Marines and Amphibious War: Its Theory and Its Practice in the Pacific*, Princeton, NJ: Princeton University Press, 1951.

Jewkes, John, David Sawers, Richard Stillerman, *The Sources of Invention*, New York: MacMillan, 1958.

Johns, Adrian, *Death of a Pirate- British Radio and the Making of the Information Age*, New York: Norton, 2011.

Jones, Neville, *The Beginning of Strategic Air Power: A History of the British Bomber Force, 1923-1939*, London, UK: Frank Cass, 1987.

Kelly, Charles J., *The Sky's the Limit: The History of the Airlines*, New York, NY: Coward-McCann, 1963.

Kennedy, Gerald F., "Quantico, VA., Marine Corps kite balloon station and base, 1917-229: Paolo L. Coletta, ed., United States Navy and Marine Corps Bases, Domestic, (Westport: Greenwood Press, 1985).

Kessner, Thomas, *The Flight of the Century: Charles Lindberg and the Rise of American Aviation*, (New York: Oxford University Press, 2010).

Kilpatrick, Carroll, *Roosevelt and Daniels: A Friendship in Politics*, Chapel Hill, NC: University of North Carolina Press, 1952.

Kintner, William R., *Forging a New Sword: A Study of the Department of Defense*, New York, NY: Harper, 1958.

Kreidberg, Marvin A., and Henry, Merton G., *History of Military Mobilization in the United States Army, 1775-1945*, Washington, DC: Department of the Army, 1955.

Larson, Henrietta M., Knowlton, Evelyn H., and Popple, Charles, *New Horizons, 1925-1950, History of Standard Oil Company*, (New Jersey) New York, NY: Harper and Row, 1971.

Leach, Paul R., *That Man Dawes*, Chicago, IL: Reilly and Lee, 1930.

Lécuyer, Christopher, *Making Silicon Valley: Innovation and Growth of High Tech, 1930-1970*, Cambridge, MIT Press, 2006.

Lejeune, John A., *The Reminiscence of a Marine*, Philadelphia, PA: Dorrance and Co., 1930.

Lessing, Lawrence, *Man of High Fidelity: A Biography*, Philadelphia: J.B. Lippincott, 1956.

Levine, Robert H., *The Politics of American Naval Armament, 1930-1938*, New York, NY: Garland, 1988.

Lewis, Tom, *Empire of the Air: The Men Who Made Radio*, New York: Edward Burlingame, 1991.

Lotchin, Roger W., *Fortress California, 1910-1961*, New York, NY: Oxford University Press, 1992.

Love, Robert William, ed., *The Chiefs of Naval Operations*, Annapolis, MD: Naval Institute Press, 1980, John Major, "William Daniel Leahy, 3 January 1937, 1 August, 1939."

Lowitt, Richard, *George W. Norris: The Persistence of a progressive, 1913-1933*, Urbana, IL: University of Illinois, 1971.

Maclaurin, W. Rupert, *Invention and Innovation in the Radio Industry*, New York, NY: Mac-Millan, 1949.

Margulies, Herbert F., *Senator Lenroot of Wisconsin: A Political Biography*, 1900-1929, Columbia, MO: University of Missouri Press, 1977.

Marder, Arthur J., *Old Friends: New Enemies: The Royal Navy and the Imperial Japanese Navy, Strategic Illusions, 1936-1941*. Oxford: Clarendon, 1981.

Marolda, Edward J., ed., *FDR and the U.S. Navy*, New York, NY: St. Martins, 1998.

McBride, William M., *Technological Change and the United States Navy, 1865-1945*, Baltimore, MD: Johns Hopkins University Press, 1998.

McDowell, Carl E., and Gibbs, Helen M., *Ocean Transportation*, New York, NY: McGraw-Hill, 1954.

Meulen, Jacob A. Vander, *The Politics of Aircraft: Building an American Military Industry*, Lawrence, KS: University Press of Kansas, 1991.

Miller, Edward S., *War Plan Orange*, Annapolis, MD: Naval Institute Press, 1991.

Mills, Geofrey T., and Rockoff, Hugh, ed., *The Sinews of War*, Ames, IA: Iowa State University Press, 1993.

Monsarrat, John, *Angel on the Yardarm: The Beginnings of the Fleet Radar Defense and the Kami Kaze Threat*, Newport: Naval War College Press, 1985.

Morison, Samuel E., Commanger, Henry Steele, and Leuchtenburg, William, *The Growth of the American Republic*, New York, NY: Oxford University Press, 1980.

Morrison, Joseph L., *Josephus Daniels – the Small-d Democrat*, Chapel Hill, NC: University of North Carolina Press, 1996.

Nash, Gerald D., *United States Oil Policy 1890-1964*, Pittsburgh, PA: University of Pittsburgh Press, 1968.

Nebeker, Frederic, *Dawn of the Electronic Age: Electrical Technologies in Shaping the Modern World, 1914 to 1945*, New York: John Wiley, 2009.

Nixon, Edgar B., *Franklin D. Roosevelt and Foreign Affairs, March 1934-August 1935*, Cambridge, MA: Belknap Press, 1969.

Nofi, Albert A., *To Train the Fleet for War: The U.S. Navy Fleet Problems,* Newport: Naval War College Press, 2010.

Noggle, Burt, *Teapot Dome: Oil and Politics in the 1920's*, New York, NY: Norton, 1965.

O'Connell, Robert L., *Sacred Vessels: The Cult of the Battleship and the Rise of the U.S. Navy,* New York, NY: Oxford University Press, 1991.

Owen, Robert L., *Remarkable Experiences of H.F. Sinclair with his Government*, (Tulsa: H.H. Rogers, 1929).

Peattie, Mark, "Japan: The Teikoku Kaigun," Vincent P. O'Hara, W. David Dickson, Richard Worth, ed. *On Seas Contested: The Seven Great Navies of the Second World War*, Annapolis: Naval Institute Press, 2000).

Perez, Robert C., and Willett, Edward F., *The Will to Win: A Biography of Ferdinand Eberstadt,* Westport, CT: Greenwood, 1989.

Phelps, Stephen, *The Tizard Mission: The Top Secret that Changed the Course of World War II*, Yardly, West Home, 2011.

Poolman, Kenneth, *The Winning Edge: Naval Technology in Action, 1939-1945*, Annapolis: Naval Institute Press, 1997.

Popham, Hugh, *Into Wind: A History of British Flying*, London, UK: Hamish Hamilton, 1969.

Postan, M.M., *British War Production*, London: Longmans, Green, 1952.

Potter, E.B., and Nimitz, Chester W. ed., *Sea Power: A Naval History*, Englewood Cliffs, NJ: Prentice-Hall, 1960.

Potter, E.B., "The Command Personality: Some American Naval Officers of World War II," William Geffen, ed., *Command and Commanders in Modern Warfare*: Proceedings of the second military history symposium, 2-3 May, 1968, (Colorado Springs: United States Air Force Academy, 1969).

Puleston, W.D., *The Influence of Sea Power in World War II*, New Haven: Yale University Press, 1947.

Ranft, Bryan, ed., *Technical Change and British Naval Policy, 1860-1939*, London, UK: Hodder and Stoughton, 1977.

Ratner, Sidney, *Taxation and Democracy in America*, New York, NY: John Wiley, 1942.

Reilly Jr., John C., *United States Navy Destroyers of World War II*, Poole: Blandford, 1983.

Reserve Officers of Public Affairs Unit 4-1, *The Marine Corps Reserve: A History*, Washington, DC: 1996.

Ries, John C., *The Management of Defense: Organization and Control of the U.S. Armed Services*, Baltimore, MD: Johns Hopkins University Press, 1964.

Rosen, Philip T., *The Modern Stentors: Radio Broadcast and the Federal Government, 1920-1934*, (Westport: Greenwood, 1980).

Roskill, Steven , *Naval Policy Between the Wars: the Period of Anglo-American Antagonism, 1919-1929*, New York, NY: Walker, 1969.

Rowe, A.D., *One Story of Radar*, Cambridge, Cambridge University Press, 1948.

Roukis, George S., *The American Maritime Industry: Problems and Prospects*, Hempstead, NY: Hofstra University Yearbook of Business, 1983.

Schlaiffer, Robert, and Heron, S.D., *The Development of Airplane Engines and Fuels*, Cambridge, MA: Harvard University Press, 1950.

Schumacher, Alice Clink, *Hiram Percy Maxim: Father of Amateur Radio, Car Builder and Inventor*, Greenville, The Ham Radio Publishing Group, 1070.

Settel, Irving, *A Pictorial History of Radio*, New York: Grussett and Dunlap. 1967.

Sloan, Alfred P., *My Years with General Motors*, Garden City, NY: Doubleday, 1963.

Sixsmith, E.K.G., *Douglas Haig*, London, UK: Weidenfeld and Nicolson, 1976.

Sobel, Robert, *Coolidge: An American Enigma*, Washington, DC: Regnery Publishing Co., 1998.

Smith, Henry Ladd, *Airways: The History of Commercial Aviation in the United States*, New York, NY: Alfred A. Knopf, 1942.

Smith, Roe Smith, ed. *Military Enterprise and Technological Change: Perspectives on the American Experience*, Cambridge: MIT Press, 1985.

Sparrow, Bartholomew H., *From the Outside In: World War II and the American State*, Princeton, NJ: Princeton University Press, 1969.

Spector, Ronald, *At War at Sea: Sailors and Naval Combat in the Twentieth Century*, London, UK: Penguin, 2001.

Spencer, P.C., *Oil and Independence: The Story of the Sinclair Oil Corporation*, (New York: Newcomen Society of North America, 1957).

Standley, William H., and Ageton, Arthur A., *Admiral Ambassador to Russia*, Chicago, IL: Henry Regnery Co., 1955.

Stein, Harold, ed., *American Civil-Military Decisions: A Book of Case Studies*, Tuscaloosa, AL: University of Alabama Press, 1963.

Stratton, David H., *Tempest over Teapot Dome: The History of Albert Fall*, Norman, OK: University of Oklahoma, 1998.

Stockfisch, Jacob A., *Plowshares into Swords: Managing the American Defense Establishment*, New York, NY: Mason Lipscomb, 1973.

Sturmey, S.G., *The Economic Development of Radio*, London, Gerald Duckworth, 1958.

Swords, S.S, *Technical History of the Beginnings of Radar*, London: Peter Peregrinus, 1986.

Taylor, A. Hoyt, *Radio Reminiscences- A Half Century*, Washington: Office of Naval Research, 1948.

Taylor, A. Hoyt, *The First Twenty-five Years of the Naval Research Laboratory*, Washington: U.S. Navy Department, 1948.

Thatcher, Harold W., *Planning for Industrial Mobilization 1920-1940*, Washington, DC: Office of the Quartermaster General, 1943.

Thruelsen, Richard, *The Grumman Story*, New York, NY: Praeger, 1976.

Tritten, James J., Introduction of Aircraft Carriers into the Royal Navy: Lessons in the Development of Naval Doctrine," *Naval Doctrine Command*, Norfolk, VA: 1994.

Tucker, Garland S., *The High Tide of American Conservatism: Davis, Coolidge and the 1924 Election*, (Austin: Emerald Book Co., 2010).

Van der Linden, F. Robert, *The Boeing 247: The First Modern Airliner*, Seattle, WA: University of Washington Press, 1991.

Vat, Dan Vander, *The Pacific Campaign: World War II, the U.S. Japanese Naval War, 1944-1945*, New York, NY: Simon Schuster, 1991.

Venable, John D., *Out of the Shadow: The Story of Charles Edison*, East Orange, NJ: Charles Edison Fund, 1981.

Wagoner, Harless D., *The U.S. Machine Tool Industry from 1900-1950*, Cambridge, MA: MIT Press, 1968.

Watson-Watt, Robert, *The Pulse of Radar: The Autobiography of Sir Watson-Watt*, New York: Dial, 1959.

Wheeler, Gerald E., *Admiral William Veazie Pratt: A Sailor's Life*, Washington, DC: Department of Navy, 1974.

White, Leonard D., *The Republican Era: 1869-1901, A Study of Administrative History*, New York, NY: Macmillan, 1958.

Wildenberg, Thomas, *All the Factors of Victory: Admiral Joseph Mason Reeves and the Origen of Carrier Airpower*, (Washington, DC: Brassey's, 2003).

Wilson, John R.M., *Herbert Hoover and the Armed Forces*, New York, NY: Garland, 1993.

Woodbury, David Oakes, *Battle Fronts of Industry*, New York, NY: J. Wiley, 1948.

Yergin, Daniel, *The Prize: The Epic Quest for Oil, Money and Power*, New York, NY: Pocket Books, 1993.

Zimmerman, David, *Top Secret Exchange: The Tizard Mission and The Scientific War*, Montreal: McGill-Queens University Press, 1996.

ARTICLES

Beers, Henry P., "The Development of the Office of Chief of Naval Affairs," Part IV, 11, *Military Affairs*, 6, No. 3 (Winter 1947).

Brainard, E.H., "Marine Corps Aviation," *Marine Corps Gazette*, 13, No. 3 (March 1928).

Dyer, George C., "Naval Amphibious Landmarks," 92, No. 8, *U.S. Naval Institute Proceedings*, August, 1966.

Ford, Harvey S., "Walter Folger Brown," *Northwest Ohio Quarterly*, 31, No. 2 (Summer 1954).

Fullam, W.F. Rear Admiral, "The Passing of Sea Power," *McClures*, 26. No. 1, (June 1923).

Gretton, Peter, (*reviewed by*) "Naval Policy Between the Wars," *Journal of the Royal United Service Institution*, 113, No. 539 (Feb/March 1968).

Heinl, R.D., "Naval Gun Support," *Military Review*, 26, No. 12 (December 1946):

Hessler, William H., "The Battleship paid Dividends," U*nited States Naval Institute Proceedings*, 72, No. 9 (September, 1946).

Hoffman, Frank, "The Myth of War Plan Orange: Lessons for Naval Innovation," *Strategic Review*, 28 (Summer, 2000).

King, Ernest J., "United States at War," *Final Official Report to the Secretary of the Navy, March 1, 1945 to October 1945*, U.S. Naval Institute Proceedings, 72, No. 1 (January 1946):

Lautenschlager, Karl, "Technology and the Evolution of Naval Warfare," *International Security*, 8, No. 2 (Fall 1983).

Lee, David D., "Herbert Hoover and the Development of Commercial Aviation, 1921-1926," *Business History Review*, 58, No. 1 (Spring 1984).

Madden, Robert B., "The Bureau of Ships and its E.D. Officers," *Journal of the American Society of Naval Engineers*, 66, (February, 1954).

MaGruder, T.P., "The Navy and the Economy," *Saturday Evening Post*, 199, (September 24, 1927).

McKinney, James B., "Radar: A Case history of an Invention," *IEEE A & E Systems Magazine*, (21, 8, August, 2006).

Morison, Elting E., "Naval Administration in the United States," *U.S. Naval Institute Proceedings*, 72, No. 10 (October 1946).

Mosley, F.M., "The Navy's Floating Dry Dock," *Marine Engineering and Shipping Review*, 50, No. 9 (September 1945):

Saxon, Timothy D., "Anglo-Japanese Naval Cooperation," *Naval War College Review*, 53, No. 4 (Winter, 2001).

"Scientific Management in the Navy," *The Literary Digest*, 39, Nov. 10 (April 10, 1912).

Seely, Victor D., "Boeing's Pacesetting 247," *American Aviation Historical Society Journal*, 9, No. 4 (Winter, 1964).

Smith, H.S., "The Development of Amphibious Tactics in U.S. Navy," *Marine Corps Gazette*, 29, No. 10 (October 1946).

Snedeker, James, "The Naval Reserve Officer," *U.S. Naval Institute Proceedings*, 72, No. 7 (July 1946).

Spruance, R.A., Admiral, "The Victory in the Pacific," 91, *Royal United Service Institution Journal*, (November 1946).

DISSERTATIONS

Alderfer, H.F., "The Personality of Warren G. Harding," Ph.D. Dissertation, Syracuse University, 1935.

Atwater, William Felix, "United States Army and Navy Development of Joint Operations, 1898-1942," Ph.D. Dissertation, Duke University, 1986.

Bowman, Chester Grainger, "The Philadelphia naval Aircraft Factory and the Emerging American Aircraft Industry, 1917-1927," Senior Thesis, Princeton University, June, 1969.

Campbell, Mark A., "The Influence of Airpower upon the Evolution of Battle Doctrine in the U.S. navy, 1922-1941," Master's Thesis, University of Massachusetts, Boston, 1992.

Christman, Calvin Lee, "Ferdinand Eberstadt and Economic Mobilization of War," *1941-1943*, Ph.D. Dissertation, Ohio State University, 1971.

Enders, Calvin W., "The Vincent Navy," Ph.D. Dissertation, Michigan State University, 1970.

Infusimo, Frank J., "The United States Marine Corps and War Planning (1900-1941)" M.A. Thesis, California State University, 1973.

Gamble, Raymond C., "Decline of the Dreadnought: Britain and the Washington Naval Conference," 1921-1922, Ph.D. Dissertation, University of Massachusetts, 1993.

Ginther, James Anthony, and McCutcheon, Keith Barr, "Integrating Aviation Into the United States Marine Corps," Ph.D. Dissertation, Texas Tech University, 1999.

Goldman, Emily Oppenheimer, "The Washington Treaty System: Arms Racing and Arms Control in the Inter-War period," Ph.D. Dissertation, Stanford University, 1989.

Henry, James Joseph, "A Historical Review of the Development of Doctrine for Command Relationships in Amphibious Warfare," Master's Thesis, U.S. Army Command and General Staff College, Ft. Leavenworth, KS: 2000.

Jenkins, Innis LaRoche, "Josephus Daniels and the Navy: 1913-1916, A Study in Military Administration," Ph.D. Dissertation, University of Maryland, 1960.

Johnson, Glenn Allister, "Secretary of Commerce Herbert Hoover: The First Regulator of American Broadcasting, 1921-1928: Ph.D. Dissertation, University of Iowa.

Keith, Francis Lovell, "United States Navy Task Force Evolution: An analysis of United States Fleet Problems, 1931-1934," Master's thesis, University of Maryland, 1974.

Madaras, Lawrence, "The Public Career of Theodore Roosevelt Jr.," Ph.D. Dissertation, New York University, 1964.

McBride, William Michael, "The Rise and Fall of Strategic Technology: The American Battleship from Santiago Bay to Pearl Harbor: 1898-1941," Ph.D. Dissertation, Johns Hopkins University 1989.

Miller, Philip Barlow, "The Role of Secretary of the Navy Edwin Denby and the Teapot Dome Affair," Honors Thesis, University of Illinois, June, 1957.

Moy, Timothy David. "Hitting the Beaches and Bombing the Cities: Doctrine and Technology for Two New Militaries, 1920-1940," Ph.D. Dissertation, University of California, Berkeley, 1992.

Reynolds, Charles V., "America and a Two Ocean Navy, 1933-1941," Ph.D. Dissertation, Boston University, 1978.

Smith, William Young, "The Search for National Security Planning," 1900-1947, Ph.D. Dissertation, Harvard University, 1960.

Svonavec, Stephen Charles, "Congress and the Navy: The Development of Naval Policy, 1913-1947," Ph.D. Dissertation, Texas A&M University, 2000.

Walterman, Thomas Worth, "Airpower and Private Enterprise Federal-Industrial Relations in the Aeronautics Field, 1918-1926," Ph.D. Dissertation, Washington University, 1970.

Wadle, Ryan David, "The United States Navy Fleet Problems and the Development of Carrier Aviation, 1923-1933," Master Thesis, Texas A&M University, 2005.

Welton Winans II, "Harry F. Sinclair and the Teapot Dome Scandal: Appearances and Realities Master's Thesis, Indiana University, 1961).

West, Michael A., "Laying the Foundation: The Naval Affairs Committee and the Construction of the Treaty Navy, 1926-1934," Ph.D. Dissertation, Ohio State University, 1980.

Wheeler, Gerald Everett, "Japan's Influence on American Naval Policies, 1922-1931," Ph. D. Dissertation, University of California, 1954.

Wilson, Desmond P., "Evolution of the Attack Aircraft Carrier: A Case Study in Technology and Strategy," Ph.D. Dissertation, Massachusetts Institute of Technology, 1965.

Wolters, Timothy Scott, "Managing a Sea of Information: Shipboard Command and Control in the United States Navy, 1899-1945), Ph.D. Thesis. MIT, 2003.

OTHERS

"History of Navy Fighter Direction," Part III, 6, Combat Information Center, (June 1946).

Historical Collections of the Great Lakes San Cristobal, Barnes-Duluth Shipbuilding Company, Duluth, MN, http://besu.edu-bin/xus 12.c91.

Leopold, Richard, "Fleet Organization, 1919-1941," *Navy Department Library*, Washington, DC: 1945), mimeo.

Mowry, George E., *Landing Craft and the War Production Board, April 1942 to May 1944*, Washington, DC: War Production Board, 1946.

Visser's Esso UK Tanker's site, http?www.aukevisser.nl/uk/inees.htm.

The Manitowoc Herald Times, November 11, 1943, Manitowoc Public Library, Manitowoc, Wisconsin.

Tillman, Barret, "Coaching the Fighters," U.S. Naval Institute Proceedings, January 1980.

Index